London's Lost Route to Basingstoke

The Story of The Basingstoke Canal

Also by P.A.L. Vine

London's Lost Route to the Sea (1965) David & Charles
London's Lost Route to Basingstoke (1968) David & Charles and A.M. Kelley, New York
The Royal Military Canal (1972) David & Charles
Magdala (1973) E.T.O., Addis Ababa
Ethiopia (1974 limited edition)
Introduction to Our Canal Population: George Smith (1974) E.P. Publishing
Pleasure Boating in the Victorian Era (1983) Phillimore
West Sussex Waterways (1985) Middleton Press
Surrey Waterways (1987) Middleton Press
Kent & East Sussex Waterways (1989) Middleton Press
Hampshire Waterways (1990) Middleton Press

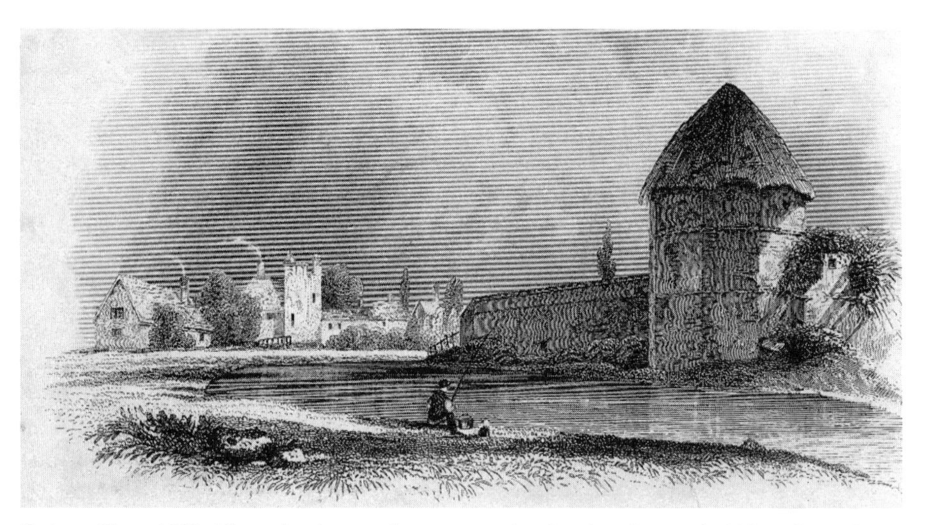

Basing village, 1838. The only nineteenth-century print showing the canal. Although canal traffic was at its busiest in the 1830s, the engraver chose to depict an angler rather than a barge. The octagonal dovecote formed part of the outer wall of Basing House

LONDON'S LOST ROUTE TO BASINGSTOKE

THE STORY OF THE BASINGSTOKE CANAL

P. A. L. VINE

SUTTON PUBLISHING

First published in the United Kingdom in 1968 by David & Charles

First published in this revised and expanded edition in 1994
Alan Sutton Publishing Limited
Phoenix Mill · Far Thrupp · Stroud · Gloucestershire

First published in the United States of America in 1968 by Augustus M. Kelley
First published in this revised and expanded edition in 1994
Alan Sutton Publishing Inc · 83 Washington Street · Dover · NH 03820

British Library Cataloguing-in-Publication Data
A catalogue record for this book is available from the British Library

ISBN 0-7509-0228-0

Library of Congress Cataloging-in-Publication Data applied for

Title page illustration: Seal of the Basingstoke Canal Navigation Company

Typeset in 11/12 Bembo.
Typesetting and origination by
Alan Sutton Publishing Limited.
Printed in Great Britain by
WBC, Bridgend, Mid Glam.

To

DR ROBERT BLAND

man-midwife and Chairman of the Company of
Proprietors of the Basingstoke Canal Navigation
1796–1816

Pleasure boaters by the swing-bridge at Zephon Common, Crookham, *c.* 1910. The wooden bridge was destroyed by a Canadian tank during the invasion scare in 1940, rebuilt in steel by Hampshire County Council in 1951, vandalized in 1987, and rebuilt and reopened in 1992

CONTENTS

ILLUSTRATIONS AND MAPS

CHRONOLOGICAL TABLE

1653	Wey Navigation opened to Guildford
1763	Wey Navigation extended to Godalming
1770	First scheme to link Basingstoke with the Thames
1778	Basingstoke Canal Act
1788	Building of the Basingstoke Canal begun
1791	Basingstoke Canal opened to Woking
1792	Basingstoke Canal opened to Pyrford
1793	Basingstoke Canal opened to Odiham
1794	Basingstoke Canal opened to Basingstoke
	Landslide at Greywell
1796	Completion of the Andover Canal
1799	Attempt to begin the branch canal to Turgis Green
1801	Bagshot Canal project
1810	Kennet & Avon Canal opened
1816	Wey & Arun Junction Canal opened
	Death of Dr Robert Bland
1818	Attempt to build the branch canal to Hook
1823	Inland water communication between London and Portsmouth established
1825–6	Berks & Hants Canal Bills presented and lost
1832	Death of Sir Richard Birnie and bankruptcy of J.R. Birnie
1838	Frimley Aqueduct built
	Peak tolls and tonnage on the Basingstoke Canal
1839	London–Basingstoke railway opened
1849	Guildford–Farnham railway opened
1854	Construction of Aldershot Camp begun
1861–2	*Basingstoke Canal Navigation* v. *Brassey & Another*
1866	Basingstoke Canal Navigation Company went into voluntary liquidation
1869	Basingstoke Canal offered for sale by private tender
1872	First collapse of Greywell Tunnel
1874	Basingstoke Canal sold to William St Aubyn
	Surrey & Hampshire Canal Company formed
1880	Surrey & Hampshire Canal Corporation formed
1883	First auction
	London & Hampshire Canal & Water Company formed
1885	Annual tonnage carried falls to 301 tons – lowest figure in the canal's history until 1948
1895	Basingstoke Canal purchased by Sir Frederick Hunt
1896	Woking, Aldershot & Basingstoke Canal Company formed

1897	Hampshire Brick & Tile Company formed
1898	Brickworks established at Up Nately
1901	Commercial traffic to Basingstoke ceased
1902	Frimley Aqueduct rebuilt
	Southampton Canal project
1904	Death of Sir Frederick Hunt
	Second auction
1905	Basingstoke Canal purchased by William Carter
1908	Brickworks at Up Nately closed
	London & South-Western Canal Company formed
1910	Last barge to Basingstoke
1911	Woking UDC succeed in High Court Action against London & South-Western Canal Company and Carter
1913	Court of Appeal set aside High Court decision
1914	Basingstoke Canal Syndicate formed
1918	Commercial traffic to Fleet ceased
1921	Commercial traffic to Aldershot ceased
1923	Basingstoke Canal purchased by A.J. Harmsworth
1932	Second collapse of Greywell Tunnel
1936	Coal traffic to Woking Gas Works ceased
1937	Weybridge, Woking & Aldershot Canal Company formed
1942	Guards' swimming pool constructed above lock XXII at Pirbright
1949	Commercial traffic on Basingstoke Canal ceased
	Third auction
	New Basingstoke Canal Company formed
1960	End of horse towage on the Wey Navigation
1964	Wey Navigation transferred to the National Trust
1966	Formation of Surrey & Hampshire Canal Society
1969	First pair of upper lock gates constructed by the Society
1973	Purchase of canal from Ash Embankment to Greywell by Hampshire County Council
1976	Purchase of canal from Woodham to Ash Embankment by Surrey County Council
1981	Frimley Aqueduct relined
1990	Basingstoke Canal Authority formed
1991	Canal reopened from Woodham to Greywell by HRH the Duke of Kent
1993	Basingstoke Canal Centre opened at Mytchett

PREFACE

The history of the Basingstoke Canal differs from that of many waterways in that while it tells of great enterprise, its commercial activities were fraught with a singular lack of success. Indeed, a glance at the map to determine the location of the Basingstoke Canal immediately provokes a host of questions as to the reason for its construction, since apparently the navigation did no more than link the market town of Basingstoke with a tributary of the Thames. Unlike the majority of waterways which were made during the latter half of the eighteenth century – the Kennet & Avon, the Leeds & Liverpool, the Thames & Severn, the Trent & Mersey – the canal would seem at first sight to have possessed little commercial justification: it joined together no network of river systems; it connected no large centre of industry with a seaport; it opened up no access to a coalfield. Why then was it determined to expend so great a sum upon such a scheme? Certainly its original promoters visualized it becoming part of a national route from London to the Bristol Channel as well as to the English Channel. In *London's Lost Route to the Sea* I told the story of how this latter link came to be built; here is its sequel, which unfolds the reason why the canal never progressed beyond Basingstoke in spite of an extraordinary number of attempts.

The canal's commercial viability was in doubt before the coming of the railways, and yet for over a hundred years every effort was made to turn it into a successful trading venture. After the original company went into liquidation in 1866, the waterway fell into the hands of a succession of speculators, few more successful than the last. Some endeavoured to revive trade, others to swindle gullible investors. However, in spite of many vicissitudes, it continued to carry barge traffic until twenty years ago, and could still without undue expenditure be developed as a waterway for recreational pursuits – a proposal which was advocated as long ago as 1911.

Today the Basingstoke Canal remains peacefully secluded among the pinewoods and heaths of the Surrey and Hampshire countryside. Truncated by the loss of its terminal point at Basingstoke, the waterway has developed a certain air of mystery, for nowadays many people are not quite sure where the canal does lead. To follow its course is to trace a forgotten artery of commerce. Certainly with perseverance you can still pole a punt as far as Greywell Tunnel, but beyond its portals lie only stalactites, darkness and the legends of the past.

I first became acquainted with the Basingstoke Canal while on army exercises during the halcyon summer days of 1947, and although there are still certain questions about the canal's history which remain unanswered, I hope that those of its secrets which I have uncovered will enlighten those whose interest and enjoyment of the waterway is increased by an appreciation of the past. The original company's records are incomplete and it is known that some papers from the offices of a Basingstoke solicitor went for salvage to help the last war effort. Yet missing ledgers, minute books and printed company reports may still remain among the discarded bundles of family documents relating to former shareholders or owners of the canal. Should anyone find them, perhaps he will let me know.

I am particularly grateful to Charles Hadfield, Harry Stevens and Alec Harmsworth for providing me with original material; I am indebted to the archivists of the Hampshire, Lincolnshire and Surrey County Councils and to the editors of the *Daily Telegraph*, the *Daily Mirror*, the *Reading Mercury* and the *Reading Evening Post* for their assistance, and to the Bodleian, the British Museum, the Institute of Civil Engineers, the Birmingham, Guildhall and House of Lords libraries, the Guildford Muniment Room and to the British Transport Historical Records Office for the research facilities they have granted me.

I must also express my appreciation to the Willis Museum at Basingstoke, the National Firm Archive, Gale & Polde Ltd., Alec Harmsworth and the Surrey & Hants Canal Society for the loan of illustrations; to Dr Roger Sellman for permission to use his maps; to J. Barton, A.B. Catling, Kenneth Clew, M. Forwood, Gerald Griffith, Adrian Hamilton, R. Harris, Kathleen Richardson, R.F.D. Sibbring, Jill Sweeney and Roy Thompson for helpful information; to Ean Begg and Karin Grunenberg for arduous assistance in exploring Greywell Tunnel; to Lucy and Olaf Lambert for their kind hospitality; to Judith Dent for typing in the evenings; and, lastly, I am indeed thankful that Diana Hanks was able to carry out much of the secretarial and some of the research work as well as being a constant source of encouragement; the reward for such painstaking work can only be found in the pleasure it may bring to others.

P.A.L.V.
SS *Queen Mary*
July 1967

PREFACE TO SECOND EDITION

Twenty-five years have passed since the first edition of this work was published and such has been the success of the Surrey & Hampshire Canal Society in restoring the waterway to something like its former self that two new chapters have been added. This change in the Basingstoke's fortunes has been due to the great enthusiasm which the Society has been able to generate among the inhabitants of every town and parish through which the waterway passes; not only have they come to watch restoration at work, but for one score years and more, groups of young, and not so young, have been out in all weathers to clear the accumulation of undergrowth and debris of the past decades. Even the hearts of tough financially-orientated county councillors have been softened by the sight of these gangs of volunteer labourers working in all seasons at what must, initially, have seemed a lost cause. The Society has not been content to labour alone. It has succeeded in attracting support from local authorities, from industry and from other waterway societies; but it has been very largely by its own efforts that it has been responsible for revitalizing a derelict canal and for creating a visually attractive waterway out of a huge overgrown ditch.

At the same time tribute also must be paid to the understanding approach of both the Hampshire and Surrey County Council authorities which initially had to be convinced, not only that ratepayers' money would be well spent, but that gangs of volunteer labour were not a one-day phenomenon that might quickly vanish once the councils had bought the canal.

This second edition has been revised and the text completely reset. Many new illustrations and plans have been included to show how the canal's appearance has changed with the passing of the years. More information has come to light about the early history of the navigation, of Pinkerton's employment of Benjamin Latrobe, who was later to build the White House in Washington, of how John Rennie came to be involved owing to William Jessop's other commitments, and of how Charles Jones was suddenly dismissed while building Greywell Tunnel. A perusal of the chapter headings and the Notes will indicate where new material has been found.

I wrote in my original preface that there were still certain questions about the canal's history which remained unanswered. Today this still remains true but the newsletter of the Surrey & Hampshire Canal Society has served as an excellent medium for inspiring people from all over Britain to write of their own recollections and experiences. However, to local historians a word of warning. Our memories for dates and details of events often become a little confused with the passage of time and a contemporary account is to be more highly regarded than one written many years after the event. Some canal histories and county guidebooks still repeat the old belief, once shared by Charles Hadfield and myself, that the canal was originally opened in 1796 whereas the opening did of course take place two years earlier, a date which only the files of a national newspaper initially corroborated. Although most of the printed reports of the company's meetings sent to shareholders between 1788 and 1830 have now been traced, surprisingly few are available for the period 1831–64. If any reader knows where any of the missing years may be found would they kindly advise me.

It is curious how many people from all walks of life have become involved in one way or the other with the canal. It has fallen into the hands of more than its fair share of speculators, like parliamentarians Sir Frederick Hunt and Horatio Bottomley, but it was its sylvan beauty which attracted men of letters like C.S. Forester and Cyril Connolly.

It is now the task of the Basingstoke Canal Authority to endeavour to meet the needs of all recreational users. I wish them well. The canal's bicentenary occurs in September 1994. I would not be surprised to learn that by the year 2094 Greywell Tunnel had been reconstructed and a branch canal built to link the waterway, if not with the Itchen Navigation, at least with the recently reopened Kennet & Avon Canal.

When I wrote a new preface for a planned 1981 revised edition which was not however published, I mentioned the names of some of the many people who had written and made helpful suggestions. They included the late Sir Eric Errington, David Gerry, Robin Higgs, Aubrey Slaughter and Jim Woolgar; I also expressed my appreciation of the work done by Rosalind and daughter Edwina who contributed to this revision with varying degrees of success. Twelve years later additional names require acknowledgement. I refer particularly to Tim Childerhouse, Hugh Compton, Mrs Mary Murphy (Institution of Civil Engineers), the staff of the Hampshire and Surrey County Record Offices and the many members of the Surrey & Hampshire Canal Society, and in particular to the Defence Research Agency, Clive Durley, Dendy Easton, Tony Harmsworth, Dieter Jebens, Stan Mellor, David Robinson and Sotheby's for providing illustrations. The difficult task of updating the earlier texts has been well accomplished by Kay Bowen.

P.A.L.V.
Pulborough
November, 1993

Conversion Tables

Imperial Measurements

1 inch	=	2.5 cm
1 foot	=	30.5 cm
1 yard	=	0.9 m
1 chain	=	20.1 m
1 furlong	=	201.2 m
1 mile	=	1.6 km

Sterling Currency Conversion

one farthing (¼*d*)	=	0.1p
one halfpenny (½*d*)	=	0.2p
one penny (1*d*)	=	0.4p
tuppence (2*d*)	=	0.8p
threepence (3*d*)	=	1.25p
sixpence (6*d*)	=	2.5p
one shilling (1*s*)	=	5p
half a crown (2*s* 6*d*)	=	12.5p
one crown (5*s*)	=	25p
half sovereign (10*s*)	=	50p
one sovereign (£1)	=	£1

The decimal system was introduced in 1971.

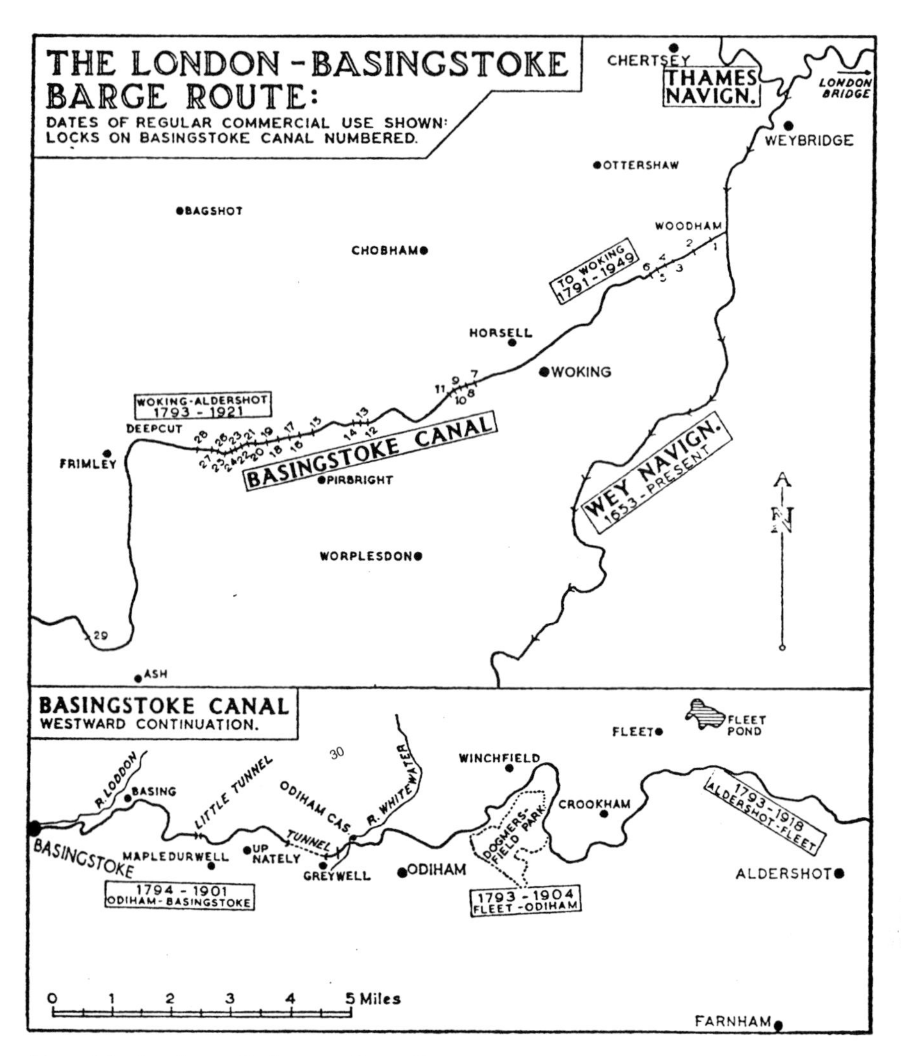

Map of the London to Basingstoke Barge Route

CHAPTER ONE

THE DEMAND FOR WATER CARRIAGE IN HAMPSHIRE

State of commerce in Hampshire in the eighteenth century – Basingstoke's importance – bad state of roads – development of waterways – success of the Wey Navigation – its extension to Godalming (1763) – demand for improved communications with London – petition for Bills to build canals from Reading to Maidenhead and from Basingstoke to Monkey Island (1770) – reasons for failure.

Eighteenth-century Hampshire had a flourishing economy. In the 1720s Daniel Defoe, in the course of one of his tours through Great Britain, particularly noted the vigorous trade of some of the market towns, especially Andover and Basingstoke. The latter, situated in the midst of woods and pastures, rich and fertile, was observed to be a large populous market town with a good corn trade. It had recently begun manufacturing 'druggets and shalloons and such slight goods which however employs a good number of the poor people and enables them to get their bread which knew not how to get it before'![1]

Defoe also commented on the great fair at Weyhill, near Andover, to which nine other counties sent their sheep, while hops came there from Kent, Surrey and Sussex, and cheese from Wiltshire, Gloucestershire and Somerset. He further remarked on the great activity in the dockyards at Portsmouth and on the revival of trade at Southampton. The corn grown in the county was generally sent to the London markets, some by Reading, some by Chertsey, but a portion of it was sent westwards to the Winchester and Newbury markets. London was also their chief market for meal, malt and timber. The textile industry still maintained its place, the new machinery being driven by water- or horse-power, and the manufacture of serges, druggets, bombazines, barragons, shalloons, tub greens and other woollen goods was fairly widespread in the county as late as 1810, before yielding soon afterwards to the onslaught of steam power.[2] Basingstoke was principally supplied with coal from Winchester, where it was generally cheaper than at London although not of such good quality; consequently supplies were also brought from Reading and Aldermaston. Iron came from Reading, Staines, Chertsey and Brentford.

In the 1830s Robert Mudie described Basingstoke as possessing

> several advantages over most of the other small towns of the country. Its situation is very beautiful; and the land about it very rich – many fields being of sufficient quality for the production of hops; and the crops upon most of the surface are of the highest order. Thus, a great deal of farm-produce – far more than is necessary for the supply of the town and neighbourhood, is raised every year, and brought to the corn-market in the town, which is well supplied and well attended. Then, Basingstoke enjoys the advantage of being a greater thorough-fare than any other town in the inland parts of Hampshire; or, indeed, in the South of England. From the South and West, the great roads through Salisbury, Andover, and Whitchurch, through Stockbridge, and through Winchester, are concentrated upon it.[3]

In the reign of William and Mary the roads in Hampshire were described by Celia Fiennes as 'very stony, narrow and steep hills or else very dirty, as in most of Sussex'.[4] The roads in Surrey were rather worse, and complaints regarding those around Godalming and Cranleigh continued into the nineteenth century. However, Hampshire fared better than most other counties in southern England in the general improvement of road transport and communications which took place after 1750. The main highways from Portsmouth and Southampton to the capital were as good as any in the kingdom and carried regular coach and waggon services. About 1750 Alton started its own coach, the 'Alton Machine', which left every morning at six o'clock and reached London the same night. Basingstoke in the 1780s was served regularly by road waggons from Exeter and Taunton, as well as by coaches from those towns and from Bristol, Salisbury and Southampton.[5] Nevertheless the cost of carriage and the condition of many of the country roads in Hampshire and Surrey limited the distribution of farm produce. In winter the state of the lanes was so bad that it was only during periods of little rain or severe frost that heavy loads of timber, coal and manure could be removed; it was also expensive to move goods by waggon for any distance, so that only the country landowners and leading tradesmen could afford the luxury of coal and imported groceries.

Rivers have been used for conveying goods from time immemorial, but it was not until the sixteenth century that legislation began to be introduced in Britain to preserve or restore the navigation of rivers naturally navigable. Although by the early part of the seventeenth century it had become possible for barges to ascend regularly above the Thames' tidal limits as far as Oxford, it was not until the Restoration that serious attempts were made to improve the country's waterways. The Speaker of the House of Commons summarized the parliamentary standpoint when he told the Lords on 2 March 1665 that

> Cosmographers do agree that this Island is incomparably furnished with pleasant Rivers, like Veins in the Natural Body, which conveys the Blood into all the Parts, whereby the whole is nourished, and made useful; but the Poet tells us, he acts best, qui miscuit utile dulci. Therefore we have prepared some Bills for making small Rivers navigable; a Thing that in other Countries hath been more experienced, and hath been found very advantageous; it easeth the People of the great Charge of Land Carriages; preserves the Highways, which are daily worn out with Waggons carrying excessive Burdens; it breeds up a Nursery of Watermen, which, upon Occasion, will prove good Seamen; and with much Facility maintain Intercourse and Communion between Cities and Countries.[6]

There was indeed a period of great parliamentary activity between 1662 and 1665, when a number of Acts for river improvement were passed. In Hampshire the need to import coal and export agricultural produce prompted Acts to make both the Itchen (1663) and the Avon (1665) navigable. In March 1663 a Bill got as far as a second reading in the Lords to make various rivers navigable to London from towns which included Farnham and Petersfield, and also from Southampton to Winchester and Alresford.

The growing importance of Guildford as an agricultural and manufacturing centre had led by 1651 to the canalization of the River Wey from the Thames at Weybridge as far as Guildford. According to Defoe the navigation was a 'mighty support to the great corn-market at Farnham', as well as serving as a central distributing point for corn, beer, paper, hoops, bark and agricultural produce, much of which went to the London markets. The timber came not only from the neighbourhood of Guildford but was also brought by 'country carriages' in summer from the woody parts of Sussex and Hampshire over 30 miles away. The opening of the extension of the navigation to Godalming in 1763 gave a strong impetus to the coal and timber trade, which was further increased by the carriage of government stores and munitions destined for Portsmouth during the American War of Independence.

The value to agriculture of water transport was becoming increasingly evident. Before navigations were developed few farmers had recourse to artificial manure, and those that did often had to fetch it laboriously by cart or packhorse from the seashore or the nearest limekiln. Their development enabled seaweed and sand to be brought inland from the coast and for limestone to be carried from the quarries to the kilns, which were built by the wharves. As a result, the cultivation of crops was improved, costs reduced and the produce more cheaply dispatched to market. The need in Hampshire was greater since it had both fewer navigable rivers and more barren heaths than most of the counties in southern England.

By the 1770s it had become apparent to the inhabitants of Basingstoke that a canal linking their town with the Thames would be of great value. Firstly, there was the need to provide some cheaper means of transporting timber to the Thameside dockyards. The *Gentleman's Magazine* drew attention to the fact that a large quantity of timber was lying unused in the countryside around Basingstoke for want of conveyance to London and the king's dockyards at Deptford, Woolwich and Chatham, where it was in great demand for shipbuilding.[7] Secondly, cheaper and better quality coal was sought; the vicinity of Basingstoke was then supplied with coal of an inferior quality from Winchester, with a little from Reading and Aldermaston, but it was contended that when the canal was constructed London coal might be supplied at 2*d* or 3*d* a bushel less than the ruling Winchester rates.[8] Thirdly, the export of cornmeal to the London market would be facilitated by obviating the necessity of carting to the wharves at Reading, Staines, Chertsey and Aldermaton.

The navigation of the Thames was at this time, however, very bad. The serpentine loops and bends of the river, the want of a convenient horse tow-path and numerous shallows occasioned much loss of time for barge traffic. The shortest distance between Reading and Monkey Island (near Maidenhead) was only 14 or 15 miles but by the river it was almost twice as far. The navigation also entailed the disadvantages of passing against a strong current and was subject to impassable floods, which often continued for several months during the winter. Phillips summed up the case for the canal as follows:

> The expense of taking a barge of one hundred and twenty tons burden from Monkey Island to Reading has been proved before a Committee of the House of Commons, to be fifty pounds, and the passage cannot be performed in less than three days, and often three weeks, and it has even been known to have been two months; whereas, by a canal, it may be performed at all seasons of the year, excepting a very hard frost, in six hours, and at the expense of only four pounds seven shillings; supposing the tonnage to be laid at a halfpenny per ton per mile.[9]

James Brindley carried out a survey of the Thames between Mortlake and Maidenhead in 1770, and while suggesting about twelve pound locks, recommended that it would be better to bypass the river by building a canal from Maidenhead to Isleworth. The proposal to build the Reading

(Sonning)–Maidenhead (Monkey Island) Canal was launched by the Corporation of Reading at a public meeting in the town hall on 9 October 1770. Two days later a meeting held in the town hall at Basingstoke supported the plan to build a 29 mile long canal from that town via Eastrop and Basing to join the proposed Reading–Maidenhead Canal near Monkey Island. Benjamin Davies who had carried out the survey in 1769[10] estimated the cost at £51,000.

In November and December petitions for a Bill to cut a canal from Reading to Monkey Island and Isleworth were supported by Abingdon, Reading and the City of London.[11] Petitions against the Bill came from numerous local landowners and inhabitants of Maidenhead, Bray and Cookham, on the grounds of 'irreparable damage and injury to the petitioners and great injuries and disadvantages to the public'.[12]

The chief objections to this scheme to improve the Thames navigation were, firstly, that this proposed canal, even if practicable, seemed to be calculated to benefit a few and prejudice many. Secondly, if the canal was built the Thames would be neglected, and in time its channels would be so choked up by diverting the water that the towns and mills depending on it would be deprived of its facilities and conveniences. Thirdly, some nearby villages would be in danger of being deluged and some of the rich land adjacent would be subject to inundation in times of flood, while above the intended canal the lands would become rushy and swampy.[13]

The parliamentary committee heard the evidence in February 1771 and refused the application for the Reading–Maidenhead Canal Bill but ordered that so much of the report as related to the Basingstoke–Monkey Island Canal should be re-committed to the committee. Thus did Brindley's plan collapse and with it the first attempt to build a waterway to Basingstoke.

CHAPTER TWO

THE BASINGSTOKE CANAL BILL (1776–8)

Link proposed with the Wey Navigation (1776) – public meetings held – petitions for and against the Bill – Lord Tylney's opposition – Act of Parliament obtained (1778) – company formed – names of original subscribers – main provisos of the Act.

It was not until 1776 that Basingstoke revised its plan for a canal and this time it decided to press for a link with the Thames via the Wey Navigation. On 9 December the *Reading Mercury* stated that the canal had 'all the appearance of public good; therefore the ground had been measured and surveyed, and the levels taken and being found very proper for the purpose'. A public meeting was held at the White Lyon, Hartford Bridge, ten days later at which it was resolved to obtain the concurrence and support of all noblemen and gentlemen interested in the project, to hold the next public meeting in April, and to open a subscription list at Messrs Hack, King & Best's at Basingstoke and at Mr Rick's in Long Acre.

The survey of the proposed line had been made by Joseph Parker and the cost estimated at £91,218; of this sum £73,918 was for making the canal, £5,300 for land and £12,000 for contingencies. Forty locks were planned with a fall of 5 ft.[14] On 1 March 1777 Charles Best, acting as clerk to the general meeting, wrote to the Earl of Portmore and Bennet Langton, the joint proprietors of the Wey Navigation, requesting their support.

Now the Wey proprietors had recently appointed George Stubbs, a Westminster solicitor, to act as their attorney following the bankruptcy of their agent, Thomas Ward, the previous year, and it was one of Stubbs's tasks to assist the proprietors to decide whether or not to allow the Basingstoke Canal to join their navigation at West Byfleet. Various estimates were made of the anticipated traffic likely to pass to and from the Wey Navigation. After taking into account the probability that the Wey would 'suffer a loss of trade of one tenth of the quantity which paid a toll of 4*s* 6*d* a ton', it seemed quite probable that no profit would accrue to the navigation if only 24,000 tons – 'of which there may be doubt' – paid a 1*s* toll for using the navigation between Weybridge and Byfleet. However, Stubbs pointed out that if 50,000 tons were carried the profit would exceed £500 p.a. and that if the canal was made direct to the Thames at an additional cost of £10,000, the Wey Navigation would stand to lose more. 'Therefore,' said Stubbs, 'it serves the interest of the proprietors to encourage the Basingstoke Canal to join their navigation rather than discourage them and cause them to drive it to Chertsey.'

At a meeting held at Hartford Bridge on 23 June 1777 to consider proper measures for carrying the canal into execution, it was estimated that boats carrying 35 tons could reach the Wey Navigation from Basingstoke in eleven hours. Tolls were calculated at £7,442 p.a. which supposed that 51 malt houses, 24 mills, 11 tan-yards and 6 brewhouses within 12 miles of Basingstoke would contribute £3,822. The timber trade was thought to be worth £600 and waggon traffic from the West Country, £500.[15]

Over a period of several months, clauses for the Bill were drafted and redrafted and on 2 February 1778 a petition 'of several noblemen, gentlemen, clergy, freeholders and others of the counties of Southampton and Surrey' was presented to the House of Commons, setting forth that the proposed canal would be of great public utility by furnishing timber for the Navy and supplying the London markets with flour and grain at a cheaper rate than they could at present.

A pamphlet entitled 'The Utility of the Intended Basingstoke Canal Navigation' distributed in 1778 stated that the canal would shorten land carriage to western counties by some 50 miles. It also mentioned that there were sixteen mills close to the line which would be able to send 'great quantities of flour to London, that the 'immense' chalk pit at Odiham contained an almost inexhaustible quantity of manure, and the 'large lime stone pit near Guildford' would provide excellent building material'.[16]

The canal was planned to run from Cooper's Meadow in Basingstoke by 'Eastrop, Basing, Maplederwell, Andwell, Nately, Skewers, Newnham, Rotherwick, Hartley, Westpall, Turgiss, Heckfield, Greywell, Odiham, Winchfield, Dogmersfield, Crondall, Yately, Aldershot, Ash, Worplesdon, Pirbright, Woking and Horsell' to join the Wey Navigation at West Byfleet, a total distance of 44 miles. A short branch (1¼ miles) was also intended from the south-east side of the turnpike road near Turgis Green through Hartley Westpall to join the canal at Hulls Farm.

It took only three weeks for the parliamentary committee to report that they had examined the petition, and found that it complied with standing orders. Rather surprisingly, only two landowners apparently dissented to the scheme.[17] The witnesses called to prove the allegations included surveyor Joseph Parker who, when asked whether the water needed for the canal would interfere with that supplied to the Itchen Navigation to Southampton, replied that all the water which would be taken for supplying the intended canal would otherwise run into the Thames.[18] Thomas May, interrogated on whether he thought that Reading would be injured by this canal, gave the

A
LIST
OF
LAND OWNERS,
Through whoſe GROUNDS,
THE
BASINGSTOKE CANAL,
IS INTENDED TO PASS;
With the Numbers to which they refer on the Plan.

No.
1 RIGHT Honourable the Earl of Dartmouth
2 Peter Searl, Eſq.
A. His Grace the Duke of Ancaſter
3 His Grace the Duke of Bolton
4 Baſing Common Field
5 John Limbry, Eſq.
6 Mr. Henwood
7 Mr. May
8 Maplederwell Common Field
9 Mr. Sumner
10 Mr. Slawter
11 Mr. Anderſon
12 Mrs. Pitman
13 { Rev. Dr. Richmond, Glebe / Mr. Small / Mr. Hall }
14 Lovelace Bigg, Eſq.
15 Nately Common Field
16 W. Withers, Eſq.
17 Mrs. Ayers
18 Mrs. Groves
19 Mr. Earl
20 Mr. Burgeſs
21 The Right Hon. the Earl of Northington
22 Mr. Webb
23 Rev. Dr. Richmond
24 Mrs. Battin
25 Right Hon. Earl Tylney
26 Mr. Bird
27 Lady Hillſborough
28 Dean and Chapter of Windſor
29 Mr. William Paice
30 Mr. Charles Biggs
31 Maplederwell Charity
32 ——— Vanſtaygen, Eſq.
33 Mr. Thick
34 Mr. Woodcock
35 Mr. Lee
36 Corpus Chriſti, Oxford
37 Mr. Sheldin
38 Mrs. Bodicot
39 Rev. Mr. Ruſh
B. Barton Heath
40 Mrs. Toll
41 Mr. Sylveſter
42 Mr. Adams
43 Mr. Hewit
44 Mr. Perdew
45 Henry Seal
46 Sir Henry Pawlet St. John
47 Charity Land
48 Mr. Nickols
49 Mrs. Loe

No.
50 Cottage
C. Broad Oak Commons
51 Mr. Swann
52 Lady Beauclerk
E. Winchfield Hurſt Common
53 Mr. Smith
54 Rev. Mr. St. John
55 Mrs. Gilham
56 Lady St. John
57 Mr. Draper
58 Mr. Clark
59 Meſſ. Gunner and Stacey
60 Mr Green
61 Meſſ. May and Vickery
F. Seven Commons
62 Mr. Soen
63 Mr. Broombridge
H. Great Heath
64 Mr. Knowles
65 Mr. Prowting
66 Mrs. Lane
67 Meſſ. Munger, Roades, Ethington and May
* Heath
68 Rev. Dr. Waters
* Heath
69 Mr. Still
70 Mr Dair
71 Gen. Aſhe and —— Penruddock, Eſq.
* Heath

Collateral Cut.

1 Mr. Wm. Paice
2 Mr. Parfett
3 Mr. Carter
4 Mr. David Paice
5 Mrs. Wheeler
6 Mrs. Chace
7 Mr. John Chace
8 Mr. Hawley

Line through Grewell Hill.

a. Nately Common Field
b. W. Withers, Eſq.
c. Mrs. Pitman of Reading
d. Mr. John Platt
e. Mr. Gregory
f. Rt. Hon. Earl of Northington
g. Mr. Abraham Young
h. Mrs. Hawkins
i. Mr. John Holden
k. Mrs Toll

List of landowners, 1777. Note the number of commons and heaths which the canal was to intersect evasive reply that he did not think that it would be of any advantage to the town. Thomas Hasker, questioned as to why he thought the canal would be of public utility, claimed that the price of carriage between Hampshire and London would be reduced by two-thirds. Leave was granted to bring in a Bill which was given a first reading on 17 March 1778.

A letter published in the *Gentleman's Magazine* in May from 'Wickhamensis' poured scorn on the plan:

> The inhabitants of a little market town in Hants, where no considerable manufacture is carried on, have unaccountably conceived an idea that if a navigable canal was made 'some way or other' from there to London, they should emerge from their present obscurity; that instead of being retailers in trade, they should become wholesale dealers and capital merchants; and that their town would in a few years be a grand emporium and mart of trade for the county, superior in consequence to Winchester, and a rival at least with the port of Southampton. . . . What success it will be attended with, time will discover, but it is shrewdly suspected that in the issue the subscriber to this scheme will be in the same distressed condition with Volumnius'* favourites, the income to support the enormous expense of finishing their canal depending chiefly upon the tolls which are to arise from casual adventitious freight, brought by wagons from the western counties, from accidental falls of timber in the neighbourhood and the produce of a chalk-pit. But the subscribers to the canal from the coal-pits to Birmingham have been greatly benefited, and why not those? Credat Judaeus Apella! Non ego.

But this correspondent's sound note of warning may have been inspired perhaps by the knowledge that the canal would be prejudicial to the Itchen Navigation from Winchester to Southampton. At the preliminary parliamentary hearing it was said that 'they would not then bring so many goods to Basingstoke as they do now, because at many seasons of the year, and in particular months, coals are superior at London than at Winchester, and are of a much superior quality'. Furthermore the price of London coal at Basingstoke was likely to be reduced by about 15 per cent if water carriage was available.[19]

Three petitions were lodged against the Bill during the committee stage. Sir James Tylney Long, on behalf of the Rt Hon. Earl Tylney, who was abroad, maintained that the canal would be greatly prejudicial to his estates at

* An affluent Roman senator who encouraged those of lowly origin to adopt ways of life unsuited to their calling, thus bringing about their ruin.

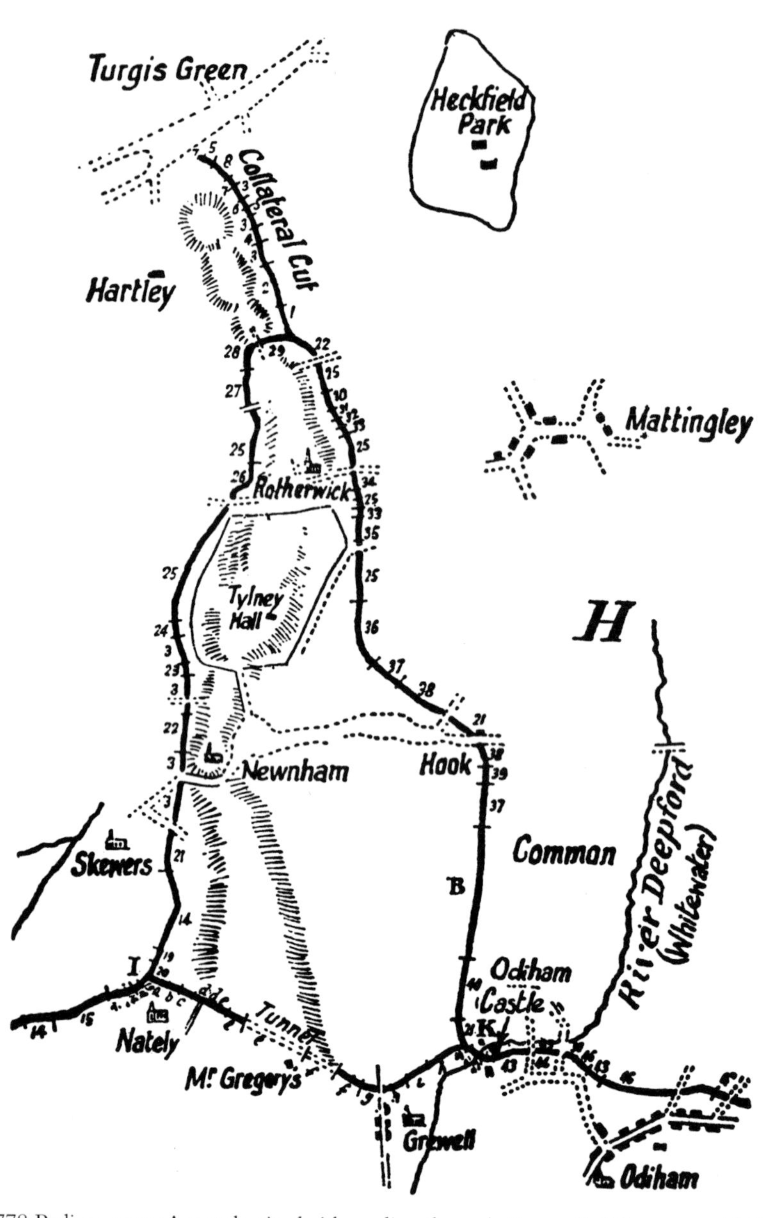

The 1778 Parliamentary Act authorized either a line through Greywell Hill or around Tylney Hall; the branch to Turgis Green was never begun

Tylney Hall. The earl's attitude was typical of many landowners up and down the country. They claimed that the water would be drawn off their land and that therefore there would not be enough left on their meadows to provide for their animals pasturing on the higher ground. They argued that the low-lying land through which the canals would naturally pass was normally the most fertile and useful for agricultural purposes, that by the digging of canals, estates would be carved up and their owners subjected to various inconveniences, and that the canals would facilitate the passage of migratory predators and thus destroy the peace and seclusion of the landlord's domain.

The second petition came from four persons who had invested money in the improvement of the Thames & Isis Navigation from London to Cricklade under an Act of Parliament passed in 1771. As the improvements had still not been completed in spite of the petitioners' having advanced large sums of money, they were opposed to the project since it threatened to divert trade between London and Reading and Aldermaston from their own navigation. Their attitude has been characteristically described by Jackman:

> It was repeatedly said that surely Parliament would not sanction one means of conveyance that would injure or destroy another which, at an earlier time, had been favoured by parliamentary authority, assistance or protection. Each navigation seemed to regard itself as the favourite child of Parliament to be jealously guarded from any adversity owing to possible or actual competition; and any upstart rival project ought to be put down so as to avoid anything that might be detrimental to property or other interests that had formerly been created under legislative sanction.[20]

The third petition came from 'several owners and occupiers of locks and winches on the River Thames below the influx of the River Kennet'.[21] These men were private individuals who had the right to levy tolls upon barges in return for the legal obligation to keep up and repair those locks and winches. Their existence was a burden upon the trading interest, but they now claimed that the construction of the Basingstoke Canal would bear heavily upon them, as much of the trade which passed through their property would be diverted, while there would be no alleviation of the tasks of maintenance and repair. There was also a petition in favour from the principal inhabitants of Farnham.

Lord Tylney's opposition had caused the promoters to hurriedly have new surveys and estimates made which showed that, by making a tunnel through Greywell Hill, the west side of Tylney Hall could be avoided and the line shortened by 6 miles, although at much greater expense. Lord Tylney thereupon withdrew his opposition to the Bill in return for an undertaking that the canal would not be built on the west side of his house without his

BASINGSTOKE CANAL,

To the RIVER WEY.

AT a publick Meeting, held at ODIHAM, on Friday the Sixteenth Day of January, 1778. in purſuance of a Reſolution entred into at the laſt Meeting,

PRESENT,

Sir SIMEON STUART, Bart. in the Chair,

Lord Rivers,
Philip Dehany, Eſq.
Joſeph Portall, Eſq.
Thomas Hall, Eſq.
John Williams, Eſq.
Alexander Baxter, Eſq.
Rev. Mr. St John,
Rev. Mr. Salter,
Rev. Mr. Prince,
Rev. Mr. Courtney,
Rev. Mr. Watkins,

Rev. Mr. Webb,
Mr. Wright, for the Earl of Dartmouth,
Mr. Kirkby, for the Earl of Portſmouth,
Mr. Haſker,
Mr. May,
Mr. Blunt,
Mr. King,
Mr. Proughton,
Mr. Williams.
Mr. Huntingford,
Mr. King, &c. &c.

The laſt Minutes read and approved.

RESOLVED, THAT the Line, which was agreed on at a Meeting held in the Town Hall Baſingſtoke on Monday the 10th day of November laſt (and which was advertized in the publick papers) be the Line adopted, namely. through the Pariſhes of Baſingſtoke, Eaſtrop, Baſing, Mappledorwell. Andwell, Nately Skewrs, Newnham, Rotherwick, Hartley-Weſtpall, Turgeſs, Heckfield, Odiham, Winchfield, Dogmersfield, Crondall, Yately, Alderſhott, Aſh, Worpelſdon, Purbright, Working, Horſell and Chertſey, to communicate with the River Wey in the ſaid Pariſh of Chertſey.

At this Meeting a Subſcription Deed was produced, read and ſigned by a Number of Noblemen and Gentlemen preſent.

RESOLVED, That the ſaid Subſcription Deed be left in the hands of Mr. Beſt, to be executed by ſuch other Noblemen, and Gentlemen, as may be Inclined to promote this uſefull undertaking.

RESOLVED, That the Petition, be preſented to Parliament, and that Mr. Beſt do wait on or write to all Perſons Intereſted, who have not already been applied to on account of the Line not having been before ſufficiently aſcertained.

RESOLVED, That the next Meeting be held at the Town Hall in Baſingſtoke, on Wedneſday the Twenty Eight Day of January Inſtant, at Twelve o'Clock.

CHARLES BEST,
Clerk to the General Meetings.

Report of the public meeting at Odiham, January 1778, confirming the line of the canal agreed at Basingstoke in November 1777. Among those present was William Wright, who was appointed resident engineer in 1788 consent,[22] and although the promoters were authorized to make the collateral branch to Turgis Green, it could not be begun until the canal had been cut from the Wey as far as Up Nately. The Bill received the royal assent on 15 May 1778, being passed through both Houses in less than three months.

The Act empowered the Basingstoke Canal Navigation Company to be formed with a capital of £86,000 in 860 shares of £100 each; to meet contingencies a further £40,000 could also be raised. No person could subscribe for more than forty shares. The original subscribers numbered thirty-three and included the Earl of Dartmouth (forty shares), the Earl of Northington, the Earl of Portsmouth (forty shares), Lord Rivers of Stratfieldsaye, Major-General Grant, and the Mayor of Basingstoke for the time being, since the Corporation had opted for five shares.

It is not difficult to deduce why these gentlemen subscribed to the undertaking. Most of them were landed gentry connected in some way or other with the counties through which the navigation was to pass. Both Dartmouth and Northington owned land at Basingstoke and Greywell, as did the Dukes of Ancaster and Bolton at Eastrop and Basing. Dartmouth was the second earl and held various important offices in Lord North's administration (1772–82), during which time he was involved in the struggle for American independence. Later, in 1786, he became high steward of Oxford University. He was also a governor of the Charterhouse, president of the London Dispensary, a vice-president of the Foundling & Lock Hospital, as well as recorder of Lichfield. He died in 1801. A plan of the canal by William Wright of Frimley in 1790 is inscribed to the earl, which suggests Dartmouth may have been involved in the planning of the waterway. The *Dictionary of National Biography* refers to him as 'an amiable, pious man who was entirely without any administrative capacity'.

Northington had become Member of Parliament for Hampshire in 1768 at the age of twenty-one. He entered the House of Lords in 1772 and in 1783 became Lord Lieutenant of Ireland. He resigned the following year with the coalition ministry and died in 1786. Portsmouth, who was the largest landowner in the county, died in 1797. Rivers, whom Walpole describes as very handsome, but a brutal, half-mad husband, became Lord Lieutenant of Hampshire in 1780. Grant, who went as a brigadier to serve under Lord Howe in America in 1776, commanded two brigades at the Battle of Brandywine the following year. He was a Scottish member of parliament for many years and became a full general in 1796. For one reason or another none of the subscribers listed in the Act was to play any significant part in the canal's history.

The first meeting of the company had to be held at the Crown Inn in Basingstoke on the first Monday in June 1778, but work could not be begun until all the capital had been subscribed and the first call of 10 per cent paid. On the other hand, if the canal was afterwards discontinued or disused for the space of five years, the lands were to be conveyed to the successors in title of those who were the owners prior to the proprietors possessing them, the consideration not to exceed the sum paid. Account books were to be kept locked up in a chest.

Optimistic landowners anticipating the discovery of coal, iron or salt mines could lay claim to these minerals if they were found during the cutting of the canal, but mines could not be worked within 20 yd of any tunnel without the consent of the proprietors. The serious consequences which could arise through working an illegal mine too close to the waterway were fully appreciated, and the company was empowered to enter upon such mines to secure them from damaging the canal at the owners' expense. The company could also make branches within 5 miles of the canal to limekilns and quarries with the consent of the landowners.

Various sections protected property owners. The canal could not pass through any land which was on 1 January 1777 'a garden, yard, park, paddock, planted walk or avenue to a house or a lawn enclosed to a mansion-house'. It could only be carried through certain parts of the estate of William Ashe and Charles Penruddocke of Woking; a bridge had to be built over Tundry Pond in Dogmersfield Park to allow carriages to pass to Sir Henry Paulet St John's 'capital house'.

Although water could be taken from any spring within 1,200 yd of the canal, none could be taken from the River Loddon or its tributaries since the mills on the river would be prejudiced, nor from any streams above the Newhaw lock which supplied water to the Wey Navigation. Neither could water be diverted from Lord Onslow's ponds, nor from those belonging to Solomon Dayrolles at Pirbright. Lord Onslow also insisted on a clause to indemnify him against any loss of revenue due to the opening of the canal since the duty he received of 4*d* a ton on Wey Navigation traffic amounted to £276 p.a.

The canal was planned to accept vessels up to 13 ft in breadth and 72 ft in length; but to conserve water, vessels loaded with less than 15 tons were not to pass through any lock without written consent. Certain conditions were imposed on the proprietors of the Wey Navigation, which supplied the only means of access to the new canal; namely a maximum toll of 1*s* per ton on all descriptions of merchandise passing over their navigation between the Basingstoke Canal and the Thames, and the requirement to keep their locks below Newhaw 81 ft long and 14 ft wide. The reason why they had to provide larger locks than those on the Basingstoke is obscure but probably relates to an agreement made earlier when the Basingstoke intended to build larger locks.

Finally the Act imposed a maximum tonnage rate on all goods and merchandise of 1½*d* per ton mile if the canal was built around Tylney Hall or 2*d* if built through Greywell. Passage and pleasure boats were to be charged 6*d* a lock. Bulk commodities like paving-stones, chalk, dung and soil could pass for not more than 1*d* per ton mile but sand and gravel for making and repairing roads were not liable for toll provided that the water was running 'through the gauge, paddle or niche of the lock'. The right to halve the duties for all types of manure instanced the regard held for the need to encourage cultivation, particularly of areas like the prodigious tracts of waste land around Bagshot Heath.

CHAPTER THREE

THE BUILDING OF THE BASINGSTOKE CANAL (1788–94)

War with America – attempts to begin construction thwarted – traffic estmates of 1786 and 1787 – capital raised – choice of engineer – William Jessop – Wright the resident engineer – Pinkerton the contractor – the sub-contractors – Benjamin Latrobe – Charles Jones – building begun (1788) – the line – chaotic management – report by John Rennie – issue of token coinage – progress – canal opened to Woking (1791) and Pirbright (1792) – additional finance required – second Act obtained (1793) – canal opened to Odiham (1793) and Basingstoke (1794).

The economic crisis which arose during the war with America precluded attempts to raise the capital for building the waterway. Speculators could not be found to invest in the enterprise and for nearly ten years the project lay dormant. Efforts were made, however, at various times to arouse interest. The peace treaty signed at Paris in September 1783 prompted the committee to publish 'An address to the Public on the Basingstoke Canal Navigation' which pointed out that 'The Company, seeing the impropriety of pushing the scheme, at a time when the nation had occasion for every resource of men and money for the expensive and complicated war in which it was then engaged, postponed all thoughts of executing it till the return of peace, which desirable event having now taken place, they propose to set about the work'. Attention was drawn to the fact that building the waterway would give employment to soldiers returning from the wars. It was considered that besides the carriage of timber, malt and flour, up traffic would also include such items as mop and prong handles, potters' ware and tanners' goods. Down traffic would be primarily coal but would also include groceries, china, glass, drapery, wine and wool. Profits were thought likely to yield at least ten per cent on the capital cost.

The end of the war was, however, followed by a severe recession and nothing further was done until November 1786, when an estimate of trade was made by an ad hoc committee of gentlemen 'who are perfectly qualified to judge the trade of that country' through which the canal was to pass. Their report contended that the cost of carriage by canal from Basingstoke to London would be 11*s* 7¾*d* a ton compared with the then actual cost of 26*s* 8*d* by land and water via Reading, or 40*s* 0*d* by waggon all the way. Tolls on the canal were expected to produce a precisely calculated annual revenue of £9,672.

The difficulty of estimating the potential traffic was considerable, for there were few data available. However, the prospective subscribers evidently doubted the accuracy of these estimates, for a more detailed report was called for and published in June 1787 which took into account local trade conditions around Basingstoke, gave a more considered view and anticipated no more than £7,783 from carrying 30,700 tons. This survey gave the chief products as coal, flour and timber, between them representing more than half the estimated traffic; the twenty-three other items listed included a variety of groceries, merchandise and household commodities. The expected return on the capital was estimated at 7½ per cent and gentlemen were invited to subscribe on the basis that until the canal was completed 5 per cent interest would be paid on each £100 share. In spite of the additional cost, the committee decided to drive the canal through Greywell Hill and avoid the circuitous route round Lord Tylney's park, thus reducing its total length to 37¼ miles. It was also agreed not to begin work on the collateral cut from Greywell to Turgis Green until the main line had been completed.

In the summer of 1787 the task of raising the authorized capital of £86,000 was begun. Between 5 November and 28 December £15,000 was subscribed. By January the total was £47,000 and at a meeting at the Crown & Anchor in the Strand on 29 February it was announced that the figure stood at £74,000.[23] A few days later the subscription list was filled with the names of some 150 prospective shareholders* and the issue of the *Reading Mercury* on 10 March 1788 mentioned that excess application money amounting to several thousand pounds had been refunded. Supporters of the Kennet & Avon Canal project drew attention to this early instance of national rather than local subscriptions since the whole sum was raised not only by the 'neighbouring gentlemen but by persons totally unconnected with that country; merchants and bankers in London and other great towns; men too well acquainted with the subject and with the value of money to engage in any undertaking which has not a reasonable prospect of advantage'.[24]

The maximum authorized subscription, £4,000, was taken up by Lord Dartmouth, Mr Newton of Litchfield, Lord Portsmouth and Lord Rivers. Other subscribers included surveyors Benjamin Davies (£300) and Benjamin Parker (£1,000); the joint proprietors of the Wey Navigation, the Earl of Portmore

* The names of the proprietors in March 1788 are given in Appendix 2.

BASINGSTOKE CANAL Navigation.

JUNE 1787.

AN ACCOUNT of the ſeveral ARTICLES which it is ſuppoſed will be carried on the intended Canal, with a Calculation of the probable Number of Tons and the Annual Produce for the Proprietors, collected and formed from a Variety of Information, and particularly from a Report recently made of the State of the Trade round Baſingſtoke.

	TONS	ANNUAL PRODUCE		
		l	s	d
Timber	5500	1411	13	4
Flour	6300	1617	0	0
Coals	6500	1668	6	8
Malt	2500	641	13	4
Grocery	1500	385	0	0
Pottery	1000	256	13	4
Paper and Materials for Paper	1000	256	13	4
Corn and Seeds	600	154	0	0
Beer	400	102	13	4
Fir Timber, Deals, &c.	500	128	6	8
Bark	1000	256	13	4
Bar Iron and Ironmongery	400	102	13	4
Cheeſe, Bacon, and Butter	400	102	13	4
Tan Yards	300	77	0	0
Hops	300	77	0	0
Wool	200	51	6	8
Hoops, &c.	200	51	6	8
Stone and Marble	200	51	6	8
Oils, Colours, Pitch and Tar	100	25	13	4
Mercery and Drapery Goods	150	38	10	0
Salt	100	25	13	4
Woollen Rags	100	25	13	4
Lead	100	25	13	4
Hemp and Flax	100	25	13	4
Chalk, Peat, Manure, &c. ½ Tonage	750	96	5	0
Weſt Country Goods	500	128	6	8
	30,700.	7783	8	4

Estimate of traffic, 1787. It was not until 1838 that this amount of cargo was carried. In only two other years during the canal's history, 1934 and 1935, was 30,000 tons exceeded

(£1,000) and Bennet Langton (£1,500), as well as his son George (£300). John Granger, the agent of the Wey Navigation, also subscribed £300.

Before the canal was begun, one of the leading shareholders, Alexander Baxter of Odiham, wrote to fellow proprietor, Joseph Portal of Overton to suggest that it might be possible to build a less expensive waterway. He thought locks might be avoided by the use of inclined planes and a narrower canal cut.* While in America Baxter had had his boats with their 3 to 4 ton cargoes drawn up some 140 ft at Niagara. 'They, with their load, are placed upon a cradle . . . and the cradle is hoisted to the top of the bank by a cable that is wound round a copestone placed upon the summit and wrought with hand spikes by the batteau [*sic*] men.'[25] Baxter supposed 'the expense of the apparatus must be very trifling' and suggested that such a contrivance could be used to avoid the lockage on the Basingstoke as well as the embanking at Ash and the tunnelling at Greywell. Baxter also suggested that Portal, who lived near 'my Lord Portsmouth', might care to meet a Mr Meadows who had attended the company meeting in London the previous June and had told him that he had invented a machine for operating an inclined plane. Baxter proposed that they met with Meadows 'either by passing a night in my house or eating a mutton chop in any of the neighbouring inns, Basingstoke excepted, as he had not had the small pox'. No more is heard of this proposal but Baxter and Portal continued to play a leading part in the affairs of the management committee.

William Jessop was appointed surveyor and consultant engineer.[26] Jessop, born in 1745, was one of the most eminent canal builders of his day. His father had worked under John Smeaton in the building of the Eddystone Lighthouse and, on his death in 1761, had left the guardianship of his family to Smeaton, who adopted William as his pupil and trained him as an engineer. Jessop acted as Smeaton's deputy in working on the Calder & Hebble and the Aire & Calder Navigations in Yorkshire. Later he was engaged in making the Trent navigable and building its tow-paths. His appointment as consulting engineer to the Basingstoke project must have been one of his first tasks on his return from Ireland where, until 1787, he was principal engineer to the Grand Canal Company. It was Jessop's custom, while holding a position of ultimate responsibility, to confine himself to acting as a consultant rather than as a resident engineer, and it is improbable that he did more than design the main engineering works and approve the method of their construction. Rolt draws attention to the fact that 'no engineer of his stature was ever more modest and self-effacing' and concluded that that is the reason why his name

* The Coteau du Lac cut, built by Lt Twiss of the Royal Engineers, was begun in 1779 and was probably the first canal navigation with locks to be built in America (*Journal of Transport History*, February 1971).

is not so well known as that of Brindley, Rennie, Smeaton or Telford.[27] On the other hand Jessop preferred to plan projects rather than carry out their execution.

In any event, the amount of time he devoted to the Basingstoke Canal was limited because he was also engaged in 1788 on a survey of the Sussex Ouse, and in 1789 on surveys of the Arun and the Western Rother as well as the Upper Thames between Dorchester and Lechlade, and in that same year he was also appointed chief engineer of the Cromford Canal. In 1790 he became one of the founders of the Butterly Iron Company in Derbyshire and was later responsible for building the Ellesmere Canal, which included the magnificent Pontcysyllte aqueduct.

The resident engineer was William Wright of Frimley whose main responsibility was to ensure that Jessop's instructions were clearly understood and carried out by the contractor and the sub-contractors.

It was hoped to complete the canal to Basingstoke in four years. The first call on the share capital of 10 per cent was made in June 1788 and two months later, notices appeared advertising for contractors to build the canal and directing interested applicants to Mr Best, the clerk, for particulars, plans and sections. Each contractor was required to bring a proper recommendation from his last employer and to find sufficient security.

The main contract was awarded to John Pinkerton and signed on 3 October 1788. Pinkertons was the largest and best known firm of early canal contractors. The firm was run by James and John, and James's sons, Francis and George. James Pinkerton's first major work had been to assist with the building of the Driffield Navigation, opened in 1770. William Jessop employed both James and John on the Selby Canal opened in 1778. Numerous undertakings followed, including the Erewash Canal (1777–9) and Dudley Canal (1785).

Pinkertons, which had to complete the undertaking within four years, was responsible for carrying out the works designed by Jessop, and at least two of the family, Francis and James, went to live at Odiham to supervise the work. Their biggest task was to organize the labour force. The usual practice was to let out the work to various sub-contractors, who in turn hired and paid the men. One of them was Benjamin Henry Latrobe (1764–1820) who was later to achieve fame as an architect in America. He designed the Bank of Pennsylvania in Philadelphia, Baltimore Cathedral and the south wing of the Capitol building, as well as the President's original White House (burnt down in 1814) in Washington. He built several buildings at Frimley while acting as superintendent of the Frimley–Ash section of the canal. There is no evidence to determine how long he was employed there but it was probably between 1788 and 1790. In 1793 he was reporting on the Blackwater Navigation in

Essex.[28] The only direct reference to his work was in correspondence about the 15 mile Chesapeake & Delaware Canal, of which he was appointed engineer in 1804, when Latrobe wrote, 'The whole of the district of the Basingstoke Canal on which I was employed was let to Pinkerton at 6¢ [*sic*] a yard'.[29]

Latrobe, whose family was of Huguenot descent, was born in Pennsylvania, spent his boyhood in England and his youth in Germany. He studied architecture on his return to England under Samuel Cokerell and engineering under John Smeaton. In 1790 he married Lydia Sellon, daughter of the rector of St James, Clerkenwell, Middlesex, who bore him two children but who died in childbirth in 1793. The tragic death of his wife led him to decide to emigrate. He left England on 25 November 1795 and reached New York in March 1796. That same year he was being consulted on improving the navigation of the Appomattox.

Another sub-contractor, who was employed on building Greywell Tunnel, was Charles Jones. Jones was both a mason and a miner with quite a past. In the 1770s he had worked in the Manchester area, moving from job to job, leaving

The eastern entrance to Greywell Tunnel *c.* 1870. Charles Jones, the unfortunate canal tunneller, lived in a house to the right above the tunnel from 1788 to 1793

debts unpaid and gaining a reputation for singular ineptitude if not downright dishonesty. However, in 1783 he had successfully tendered for driving the 2 mile long Sapperton Tunnel on the Thames & Severn Canal on the recommendation of Josiah Clowes, the resident engineer, as one 'well qualified by Experience to take the Conduct & Management' of the work. The Thames & Severn management committee were seemingly ignorant of his character, but it was not long before they discovered that Jones failed to pay his men regularly and that he was unreliable. By the time the committee learnt that his two sureties were not forthcoming, he had already been at work several months. Although he was arrested three times for debt and sent to gaol, he managed to avoid breaking his agreement with the company, who could dismiss him if he was absent for twenty-eight days at any time. The part of the tunnel which Jones was building made little progress; and after the roof had collapsed and he had gone on a two- or three-day drinking spree, he was given three months' notice in June 1785 to complete the work he had begun and then to quit. Jones retaliated with the threat of legal proceedings and some years later unsuccessfully filed a Bill in Chancery claiming he had built the tunnel.[30]

Nevertheless, in the autumn of 1788 he cajoled Jessop and the proprietors into giving him their tunnel job. It was not long before he fell foul of some members of the committee who, on their survey of the line in August 1789, came to see how the tunnel was progressing. The minutes of the general meeting are silent as to the exact cause of their complaint but Pinkerton with his 'privity and consent', agreed to Jones's instant dismissal on the grounds of 'improper conduct', almost certainly due to inebriation, with the proprietors' stipulating that he was never to be engaged again on any of the canal works.[31] Yet what gives food for thought is that Pinkerton did a few weeks later request the management committee to reinstate him. The proprietors would have none of it – Jones had certainly offended members of the management committee, not only at the time of their recent survey, 'but previous to it' – and in November they moved a resolution for the meeting to request particularly that the committee should 'continue to interfere in all such cases'.[32] *

As the result of one of the sub-contractors, Richard Hudson ('who lately undertook a small portion of open-cutting') absconding with the wages of his

* Charles Jones continued to reside at Greywell. His house is marked with his name on William Wright's 1790 plan of the canal as being by the East end of the tunnel. Interestingly enough John Rennie wrote to him there on 4 September 1790 in connection with his survey of the extension to Andover, 'I will be obliged to your taking a horse and riding over to Romsey and getting Mr Green to go with you to show you how the Bench Mark on the elm tree at Kitcomb Bridge stands in respect to the water level. You know the mark is on a tree on the west side of the road and is 1 ft 10 inches under the surface of the road opposite an elm tree on the east side.' Details of Jones's later work will be found in P.A.L. Vine, *London's Lost Route to Midhurst*, the story of the Rother Navigation.

'many country hands', Pinkertons found it necessary to inform the workmen that they should choose their own foreman 'and the monies earned to be paid every fortnight, according to the usual custom'.[33] At the request of the proprietors, preference was given to employing local labour, but as the Reverend Stebbing Shaw commented, 'such is the power of use over nature, that while these industrious poor are by all their efforts incapable of earning a sustenance, those who are brought from similar works, cheerfully obtain a comfortable support'. The workmen did in fact often travel round the country in the wake of the contractors. Denys Rolle, writing in September 1793 to the clerk of the Salisbury & Southampton Canal, says that 'some came with the engineers to the canal at Cirencester I saw and some with them to Basingstoke'.[34]

In December 1788 Pinkertons advertised for more sub-contractors to set up brick-fields along the line, advised that Mr Joseph Stringfellow of Basingstoke or Mr Wildgoose* of Horsell would know the places where clay could be found and offered encouragement to a 'sober steady man, accustomed to the brick- making business' to act as superintendent.[35] That same month the committee requested landowners or their agents on the line of the canal to meet at the George Inn at Odiham 'to treat with them about a proper mode of purchasing their several lands'.[36] Odiham was chosen because, of the seventy-one landowners listed, sixty-three possessed land above Crookham. Of the 23 miles below Crookham, over 20 were heathland.

The cutting of the canal was begun in October 1788 from the Wey Navigation at Woodham, some 1¾ miles south of Weybridge and 3 miles from the Thames. From this point it passed through Woking, Brookwood and Pirbright to Frimley and Ash, where it crossed the River Blackwater and entered Hampshire. Within this distance of 16 miles there was a rise of 195 ft, which required the building of twenty-nine locks large enough to admit vessels 72 ft 6 in. long and 13 ft 6 in. wide. At Pirbright the canal rose 97 ft by means of fourteen locks within 2 miles to reach the entrance to the great thousand-yard cutting – in places 70 ft deep – which came to be known as Deepcut.

Beyond Ash Wharf, a long curving embankment was built over the low-lying ground through which the River Blackwater flowed. From Aldershot Lock the canal was dug on the same level for 21 miles to its terminus at Basingstoke. Its route lay through the great heath between Aldershot and Fleet, then round the park at Dogmersfield and the borders of Winchfield to Odiham, where the canal crossed the River Whitewater by means of an

* George Wildgoose was employed as a land-surveyor for the Surrey Iron Railway between 1800 and 1803.

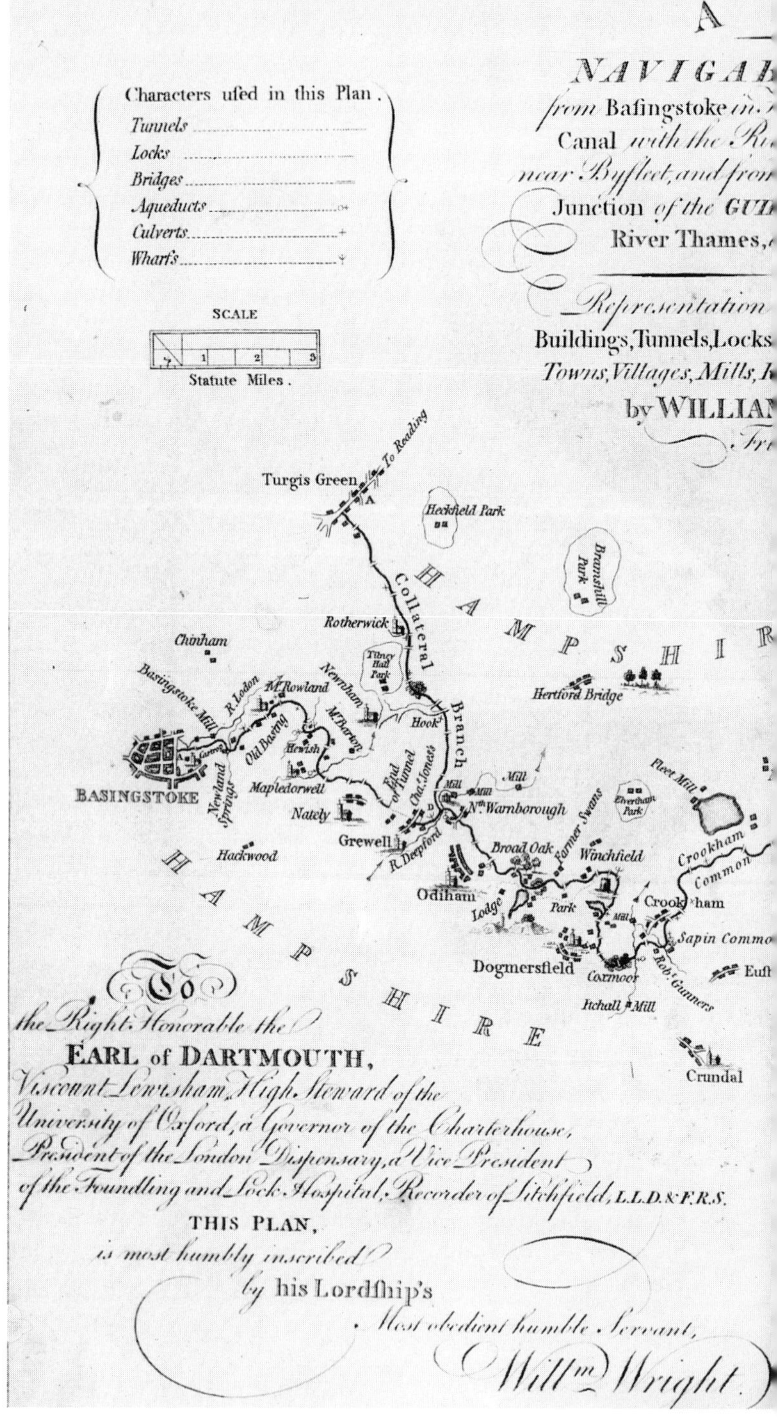

William Wright's plan of 1790 shows Greywell Tunnel and the collateral branch to Turgis Green, which the Act stipulated could not be begun until the canal had been cut to Up Nately

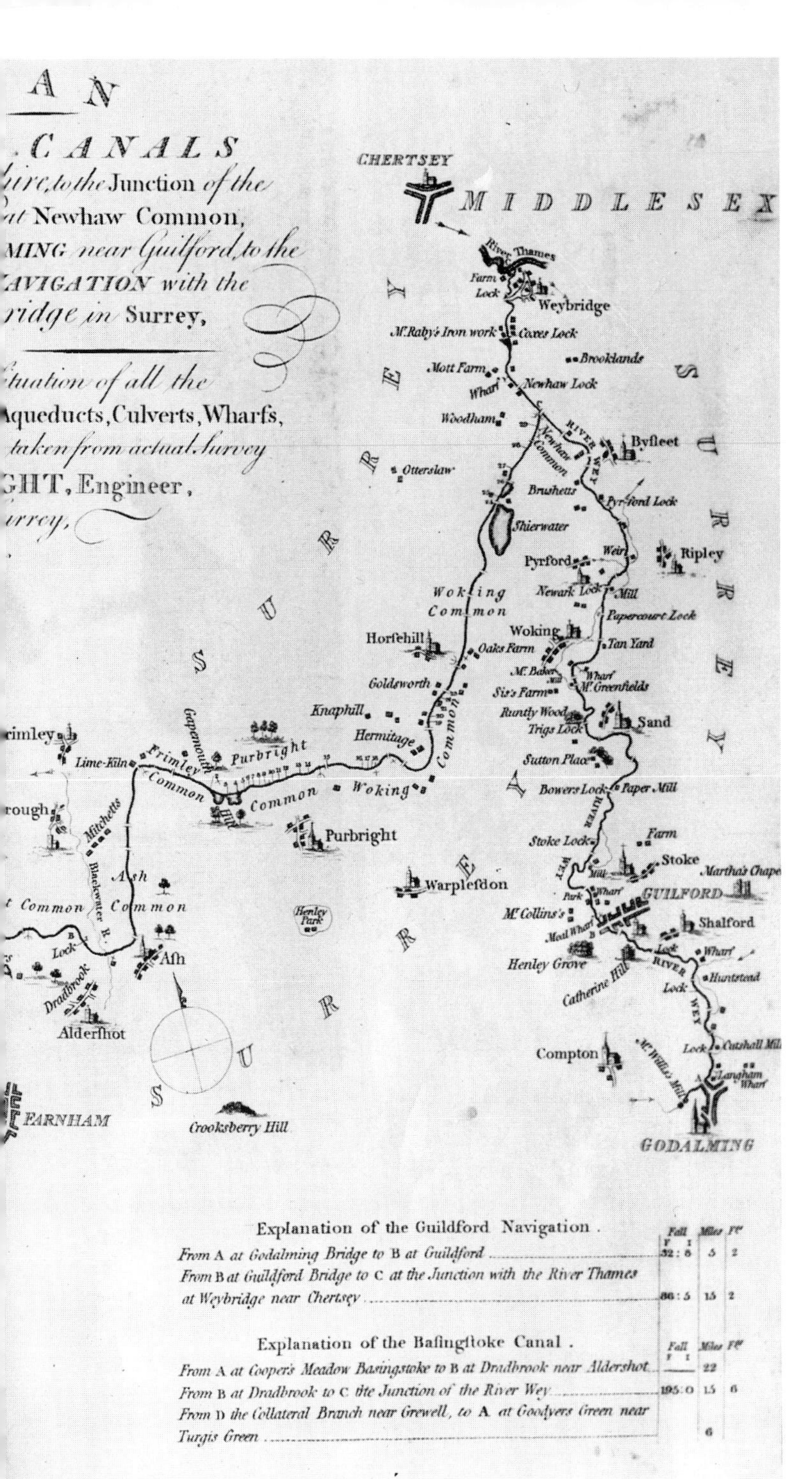
A N
CANALS
ire, to the Junction of the
at Newhaw Common,
MING near Guilford, to the
AVIGATION with the
ridge in Surrey,
tuation of all the
Aqueducts, Culverts, Wharfs,
taken from actual Survey
GHT, Engineer,
urrey,
CHERTSEY
MIDDLESEX
River Thames
Farm Lock
Weybridge
Mr. Raby's Iron work
Coxes Lock
Brooklands
Mott Farm
Wharf
Newhaw Lock
Woodham
Newhaw Common
RIVER WEY
Byfleet
Otterslaw
Brushetts
Pyrford Lock
Shierwater
Weir
Ripley
Pyrford
Newark Lock
Mill
Woking Common
Papercourt Lock
Horsehill
Woking
Tan Yard
Oaks Farm
Mr. Baker
Mill
Wharf
Mr. Greenfields
Goldsworth
Sis's Farm
Knaphill
Runtly Wood
Trigs Lock
Sand
Hermitage
Common
Sutton Place
Rimley
Lime-Kiln
Frimley Common
Gapenouth
Purbright
Common
Woking
Bowers Lock
Paper Mill
Rough
Mitchetts
Purbright
Stoke Lock
Farm
Blackwater R.
Ash Common
Warplesdon
Stoke
Mill
Martha's Chapel
Park
Wharf
GUILFORD
Common
Mr. Collins's
Henley Park
Shalford
Meal Wharf
Lock
Ash
Henley Grove
Lock
Wharf
Dradbrook
RIVER WEY
Huntstead
Catherine Hill
Lock
Aldershot
Compton
Lock
Catshall Mill
Mr. Willis's Mill
Langham Wharf
FARNHAM
Crooksberry Hill
GODALMING
SURREY
Explanation of the Guildford Navigation.
Fall F I | Miles | Fr
From A at Godalming Bridge to B at Guildford ... 32 : 6 | 5 | 2
From B at Guildford Bridge to C at the Junction with the River Thames at Weybridge near Chertsey ... 86 : 5 | 15 | 2
Explanation of the Basingstoke Canal.
Fall F I | Miles | Fr
From A at Cooper's Meadow Basingstoke to B at Dradbrook near Aldershot ... | 22 |
From B at Dradbrook to C the Junction of the River Wey ... 195 : 0 | 15 | 6
From D the Collateral Branch near Grewell, to A at Goodyers Green near Turgis Green ... | 6 |

Odiham Castle. The ruins of the fourteenth-century keep stand on the banks of the canal close to where the aqueduct crosses the River Whitewater. Its appearance in 1784, shortly before the waterway was begun, has not greatly changed

aqueduct sited close to King John's Castle. Approaching the village of Greywell a tunnel 1,230 yd long carried the canal to Up Nately. From this point it made a roundabout entry into Basingstoke through Mapledurwell and the grounds of Basing House.

Doubts about the adequacy of the water supply resulted in Jessop advising in the late summer of 1788 that the depth of the summit level should be increased by 1 ft to 5 ft 6 in. At the general meeting held in February 1789 it was agreed that John Pinkerton should receive an extra £700 over and above the sum mentioned in the contract, on condition that the canal was made watertight, this sum to be left in the hands of the company for three years after completion.

In May 1789 Jessop had had a meeting with the owner of Dogmersfield Park, Sir Henry Powlett St John Mildmay, at which Jessop agreed that instead of building a bridge over Tundry Pond as required by the Act, it was both practicable and advisable to carry the canal in the direction pointed out by Sir Henry. This was through Bath Coppice, Smith's Piddle, Mr Hall's Moor, Sir

Henry's Moor and a piece of glebe land, provided two cottages near the pond head were removed.

Because of the size of the task involved, a start had also been made in the autumn of 1788 on digging the cutting and tunnel through the chalk hill at Greywell. A few weeks after work began the Revd Shaw visited this spot and described how he saw

> above 100 men at work preparing a wide passage for the approach to the mouth, but they had not entered the hill. The morning was remarkably fine, and such an assembly of these sons of labour greatly enlivened the scene. . . . The Property under which this tunnel is intended to pass, belonged lately to Lord Northington, but now by purchase to the present Lord Dorchester. The hill cloathed with a beautiful growing of oak, called Butter-wood which united with another part, called Berkeley, extends a considerable length.[37]

The tunnel was one of the longest ever completed in the south of England.* The *Universal British Director of Trade, Commerce & Manufacture in England & Wales* (*c.* 1795) mentioned that the tunnel through Greywell Hill, 'one hundred and forty feet below the surface, is a wonderful proof of the power and art of man. A regular and constant navigation is now carried on through it and demonstrates the usefulness as well as the greatness of the undertaking.'

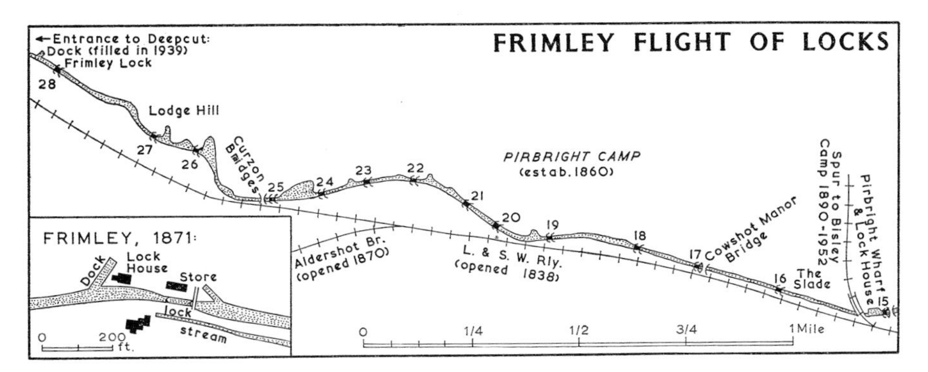

The fourteen locks between Pirbright Wharf and Frimley. Note how the London & South Western Railway is only separated from the canal by an embankment or a brick wall

* The longest was Strood (3,909 yd) on the Thames & Medway Canal. The bricks used for building Greywell Tunnel each exceeded 10 lb in weight compared with 5 lb or so for the present-day building brick.

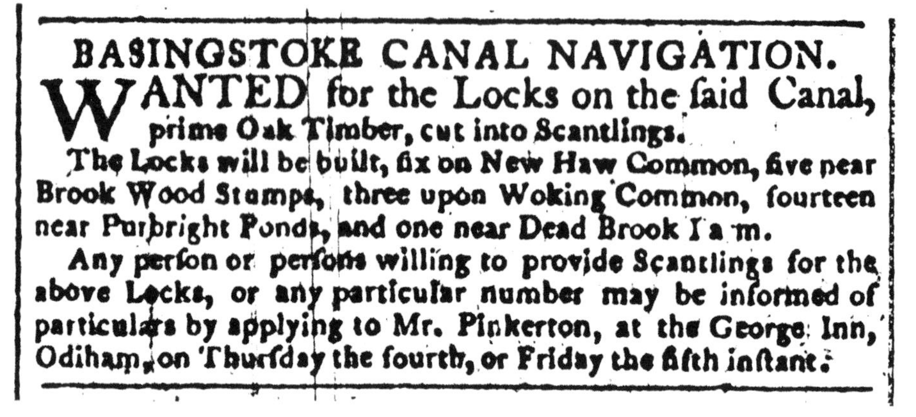

BASINGSTOKE CANAL NAVIGATION.

WANTED for the Locks on the ſaid Canal, prime Oak Timber, cut into Scantlings.

The Locks will be built, ſix on New Haw Common, five near Brook Wood Stamps, three upon Woking Common, fourteen near Purbright Ponds, and one near Dead Brook Dam.

Any perſon or perſons willing to provide Scantlings for the above Locks, or any particular number may be informed of particulars by applying to Mr. Pinkerton, at the George Inn, Odiham, on Thurſday the fourth, or Friday the fifth inſtant.

In May 1789 the company advertised in the *Hampshire Chronicle* for oak timber for building the twenty-nine locks

The 29 locks were built in several groups. The first set of 6 at Woodham, a further 5 at Woking, 3 at Brookwood and the major flight at Frimley, where 14 locks raised the canal some 100 ft in less than 2 miles. A unique feature in their construction – an example of Jessop's ingenuity – was the placing of elm beams beneath each apron behind the gates. The elm beams were placed on 3 or 4 ft of clay filling and protected by brick piling to prevent water percolation through the sandy sub-soil. When concrete was used during lock reconstruction some years ago it proved unsuccessful.[38]

The length of the line made strict supervision of the sub-contractors difficult and it is evident that the standard of building was sometimes very poor. There were complaints early in 1789 that John Pinkerton had deviated from the specifications and had ignored the warning given by the resident engineer. The committee could do no more than record that the company were 'sorry to receive such complaints of ill conduct and that they must wish for the future that Mr Pinkerton takes care that his agents strictly conform to Mr Wright's admonitions'.[39] In August it was reported that the bricks being made were 'so exceedingly bad that much the greatest part' was totally unfit for use. Mr Benjamin Davies the surveyor and Richard Harmsworth were asked to examine all the bricks along the line and provide carts for removing the faulty ones out of the way of the workmen; consequently 161,480 were condemned. In 1804 the company's surveyor reported that he would have to rebuild several of the wing walls and sides of the locks which, to the disgrace of the contractor as well as the company's then resident engineer and inspector, were put together, excepting the outer layer of bricks, with a composition of sand and rubbish 'almost without lime'.

Between June 1788 and February 1790 six different surveyors were employed. George Pinkerton received £557, Jessop, £198 and Wright, the resident surveyor, £195. The others named were Davies (£108), Mr Bull (£25) and Messrs Parker & Hodgkinson (two guineas). Although Jessop was referred to as the principal engineer, it is evident that he rarely visited the line and no mention of him is made after 1790.

Adverse comment on the proposed Andover Canal in May 1789 by 'Observator' in the *Hampshire Chronicle* referred to the futility of the Basingstoke Canal scheme, 'where the subscribers are daily selling their shares at a considerable loss, even in this early stage of the business'. The Basingstoke proprietors directed Charles Best, the clerk, to reply that this information was 'grossly unfounded, designed to serve a particular purpose and calculated to mislead the public. The word "futility" is totally misapplied to this undertaking, the success of which they continue to have every reason to entertain such sanguine expectations of, that not one single share has been sold under par, but on the contrary a premium has generally been given.' However, I doubt that this was so. William Russell of Basingstoke* wrote in July to Mr Best to state that while his £100 subscription was too trifling to be of benefit either to himself or the undertaking, he had endeavoured without success to dispose of it and hoped therefore that the proprietors would accept the £20 he had already paid as its surrender value. They did.

The great outburst of canal development and speculation in Britain had begun in 1789 and, because Jessop was the leading engineer at the time, it was natural that most promoters tried to get Jessop to carry out a survey, advise on the practicability of the scheme and help their Bills through Parliament. When he was unavailable he put forward the names of others, which was how John Rennie first came to the attention of Alexander Baxter of Odiham, the chairman of the management committee.

By the summer of 1789 the proprietors had begun to have doubts as to the efficacy of Wright and Pinkerton. Jessop, with little or no time to spare to check on their work,† suggested that Rennie should be invited down to meet the management committee. Rennie was then aged twenty-eight and at the beginning of his successful career. He expected to be away two or three weeks

* See Appendix 2.

† Jessop was soon to become chief engineer of the Cromford Canal (1789–94), the Nottingham Canal (1790–6) and the Leicester Navigation (1791–5). The first and last of these companies paid Jessop £350 p.a. on the understanding that he gave one third of his time to the job. In 1789 and 1790 he was also engaged in surveying the Ipswich & Stowmarket Navigation, the Sussex Ouse, and the Rother Navigation in West Sussex, in advising on the building of Selby road bridge and recommending improvements in lock design on the Thames, as well as appearing before three House of Lords committees on canal and navigation bills.

but in the event his visit lasted only a few days. His report to Baxter dated 28 September 1789 began 'at the request of Mr Jessop', and related how he had examined the bench marks of Mr George Pinkerton and Mr Wright along the summit level. These were unsatisfactory; Pinkerton's were

> mostly wrong, some even to the amount of several feet. Mr Wright has not established a general set of marks; there are only six to which he principally refers, many of them admitting of various constructions, viz the milestone on the Farnham road – a tree in the copse at Crookham, a post in the sand at Tundry pond, the door cill of the George Inn, Odiham, a stake in the Nately end of Jones's deep cutting and an ash [tree] in the meadow below Basingstoke – with these six marks my levels do not perfectly agree. But the difference is not sufficient to set his aside. I must therefore say so far as these marks enable me to judge, Mr Wright is adequate to the business of levelling.

He felt unnecessary to state his several other observations, as Mr Jessop would be with them by the time his letter was received. Rennie's bill, dated 6 July 1790, was for £33 8*s* 0*d*, which included the services of a land surveyor and his assistant whose task it was 'to drag the chain'.

In August 1790 Pinkerton gave public notice that the remaining part of the canal would be set out immediately after harvest and materials of all kinds would be supplied for workmen, so that a number of hands would have continued employment during the winter season. Lock carpenters and bricklayers were told they would meet with encouragement by applying to the contractors at Odiham or their agents upon the line. It is probable that a number of French prisoners-of-war assisted in the building of the tunnel and the upper reaches of the canal. A local history refers to the fact that they were incarcerated in the great chalk pit (off the Alton Road) at Odiham. There is also a reference to the employment of women as navvies at Greywell.[40]

The excavation of the basin at Basingstoke revealed a Saxon idol[41] and while the canal was being dug round Basing House, where it cut through the remains of some of the old defences, a local watchmaker discovered 800 golden guineas which were reputed to have been buried during Cromwell's siege of the house in 1645. Treasure believed to be worth more than £3 million may lie buried there and the owner was reported in 1963 as endeavouring to find it – 'the first places to be explored will be a section of the canal bank and the vaults of the house'.[42]

Large employers of labour during the late eighteenth century often found great difficulty in obtaining sufficient quantities of coin to pay their workers, due to the partial lapse of the official copper coinage during the years 1775 to 1797 when only an inadequate quantity of halfpence and farthings were made at the Royal Mint. Although token coins had been suppressed in the reign of

The shortage of copper and silver coins as a result of the war with France led to tokens being issued to the navvies for use in local shops or taverns. This 1*s* was one of those tokens issued by the main contractor, John Pinkerton, to his workmen

Charles II by a proclamation announcing that those who should 'utter farthings, halfpence or pieces of brass or other base metals, with private stamps' would be indicted, their issue became so considerable that they almost superseded the national currency. Six hundred tons of copper were used between 1787 and 1797 in the coinage of tokens in Birmingham alone and those struck in London absorbed many more tons of metal.[43]

To overcome this shortage Pinkerton, too, produced copper and silver tokens in 1789 and 1790 with which to pay the workmen. Although not issued under the authority of the Mint they differed from truck tickets in so far as they purported to be more of a general medium of exchange. They were changeable at a number of public houses including the George at Odiham. The design of the coin, believed to be the work of the celebrated Wyon, engraver to the Mint, consisted of a spade and mattock in a wheelbarrow on one side and a man and a tree trunk in a sailing barge on the reverse.

The running of the concern in its early days was chaotic. The management committee knew nothing about building canals and had apparently little business acumen. The fact that few of the subscribers were penalized for failing to pay their calls on time, instances the lax state of affairs. Not until November 1789 was any action taken and only in February of the following year did the 'gentlemen of the committee of accounts', nine in number, (including both George Stubbs and Dr Bland) draw attention to the fact that 33 subscribers had failed to pay the first call (12 August 1788), 53 the second (25 March 1789), 65 the third (1 August 1789) and 54 the fourth (9 November 1789), which had caused a shortfall of over £10,000.[39] It was therefore resolved at this late stage that shares were to be forfeited if calls were not paid within two months. At the same meeting, however, the company agreed to pay 5*s* to each member attending a committee meeting. At the next meeting, held in August, the principal concern again was the poor response to the calls for payment. Only now was the clerk ordered to ascertain whether, at a general meeting, the proprietors could legally declare shares forfeited, which had not been fully paid. Among those who had neglected to pay before 20 August 1790 was the Earl of Portsmouth, who had not paid any calls on his forty shares.

In February 1791 six members of the management committee including the Earl of Dartmouth were 'removed' because they had failed to attend meetings. These gentlemen, on reading the minutes, were not amused and at the April meeting it was hastily pointed out that their removal from the committee did not imply any disrespect and that the meeting desired to 'acknowledge their obligations to the Earl of Dartmouth and the other gentlemen not re-elected'.

In 1791 the first tolls were collected on twenty-eight tons of merchandise carried from the Wey Navigation to Horsell, which at that time was better known than the village of Woking, situated on the opposite bank of the canal. Henry Eastburn,* the resident engineer who had replaced William Wright in 1790, reported in June that 550 men and 48 horses were employed on the line besides the teams bringing materials. Three locks were in use and twelve others completed along with 223 yd of the tunnel at Gruel Hill.

By 1792 a large portion of the canal had been built, and tolls were being received on traffic up to Pirbright. Concern was however being expressed about the condition of the lower part of the Wey Navigation. In May Stubbs wrote to the Wey proprietors (the Earl of Portmore and Bennet Langton), stating that, after speaking to Joseph Nickalls, the Wey Navigation's engineer,

> he is exceedingly sorry to find his account of the defects and wants of repair so great and absolutely necessary to be done and that no directions are

* Henry Eastburn (1753–1821), a pupil of, and later assistant to, his uncle John Smeaton, was appointed on Jessop's recommendation with William Cartwright as his assistant. Both were very competent men who worked together for three years before going off to work on the Lancaster Canal. Eastburn had been elected to the Smeatonian society in 1789. In July 1792 he declined the post of joint engineer of the Horncastle & Sleaford Navigations at a salary of £300 p.a.

There is a curious reference to Cartwright in the management committee minutes of the meeting in April 1789, which stated that Mr Wright was empowered to send for William Cartwright when necessary. However, in May this resolution was rescinded and the power of sending for Mr Cartwright was rested only in the committee. Cartwright was presented with a silver cup in October 1795 'as a reward for his extra care and attention in superintending the foundations of the Lune Aqueduct on the Lancaster Canal'.

given relative thereto. The season is advancing fast and if defects and repairs are not set about immediately the season will be lost, and I need not point out of how much consequence a stoppage will be to your Lordship's interest, besides it is absolutely necessary that the river Wey and its locks should be in the best repair previous to the working of the Basingstoke Canal, because should any impediment or hindrance arise to that navigation on account of the defects of the Wey I know not what the consequences may be. As your Lordship and Mr Langton mentioned the business to me some time since I thought it my duty to say thus much that I may not in the least be blamed hereafter.[44]

General meetings were held four times a year in London and Basingstoke alternately. Those in London were well attended, 52 proprietors being present at the Crown & Anchor Tavern in the Strand in February and 53 in August 1792, compared with about twenty on average at the Town Hall or the Crown Inn in Basingstoke.

At the annual general meeting in August it was reported that about 34 miles of canal were dug, 24 of the 29 locks had been completed, 52 bridges built, 4 lock-houses finished, aqueducts constructed in the Ash Valley and over the River Deepford (Whitewater), and 884 yd of tunnel bored at Greywell. 'The two great works of Frimley Hill and the embankment in Ash Valley are nearly completed.' It was resolved that the committee should take land

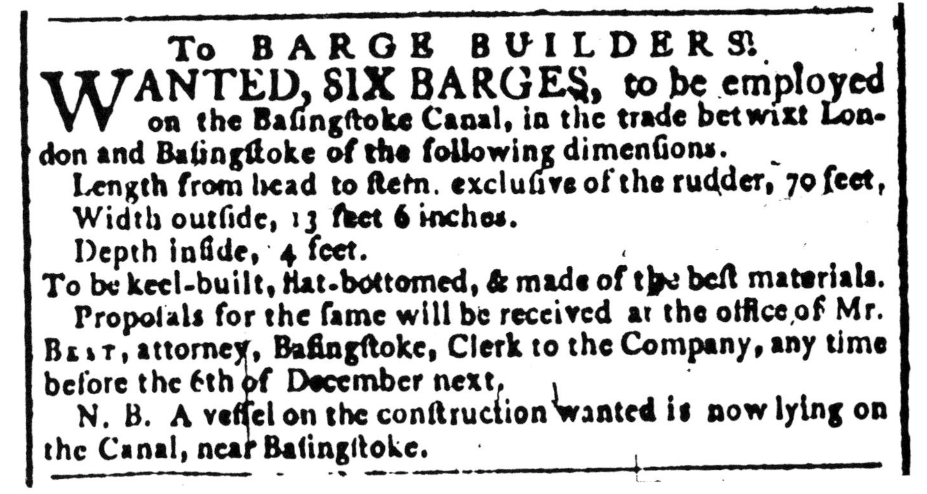

To BARGE BUILDERS.

WANTED, SIX BARGES, to be employed on the Basingstoke Canal, in the trade betwixt London and Basingstoke of the following dimensions.

Length from head to stern, exclusive of the rudder, 70 feet,

Width outside, 13 feet 6 inches.

Depth inside, 4 feet.

To be keel-built, flat-bottomed, & made of the best materials.

Proposals for the same will be received at the office of Mr. BEST, attorney, Basingstoke, Clerk to the Company, any time before the 6th of December next.

N. B. A vessel on the construction wanted is now lying on the Canal, near Basingstoke.

In November 1792 the company advertised in the *Hampshire Chronicle* for six barges. None was obtained, however, until 1794 for wharves and erect such buildings as it had been authorized to do, and that it should consider the number and kinds of staff required for the complete canal, their duties and salaries, the number and kinds of vessels to be built or engaged, and the best means of receiving rates and duties.[45] Only five locks, a few bridges, 216 yd of tunnel and 3 miles of canal remained to be completed to link London with Basingstoke. By the beginning of November only 2 furlongs remained to be begun; the committee advertised for six barges, 70 ft long from head to stern exclusive of rudder, 13 ft 6 in. wide and 4 ft deep, and stated that they were to be keel-built, flat-bottomed and made of the best materials.[46] Proposals were requested before 6 December and it was recorded that a vessel of the required construction was lying on the canal near Basingstoke.

The Act had stipulated that 5 per cent interest was to be paid annually on the sums subscribed until the canal had been completed, but no interest had been paid since 24 June 1790. Calls had been made on the share capital at varying intervals; three in 1789, three in 1790 and the tenth and last call was paid in the spring of 1792. Share certificates were issued dated 7 November 1791. Additional subscriptions for the remainder of the authorized shares were then requested and a further £39,970 was raised to bring the company's capital up to £125,970. This money had now been spent; debts were outstanding and it was evident that there were no funds to complete the work. The only recourse the company had was to obtain a second Act which was indeed granted in March 1793. The president and scholars of Magdalen College, Oxford, lodged a petition against the Bill because although 'owners and proprietors of tythes, arising, issuing and growing out of, and from, the said lands and grounds through which the said canal runs' had been 'injured to a very considerable amount, they had not been able to obtain compensation'.[47] The petition was dropped and a meeting of commissioners appointed by the Act met on 4 April in Basingstoke to settle the matter. The second Act empowered the proprietors to raise, either by loan or by annuities on mortgage of the tolls, £60,000, in addition to the £126,000 already raised, to finish the canal.

It is an astonishing fact that at this stage the completion of the canal should have been frustrated through the unimaginative management of the company's affairs, the full story of which did not come to light until 1800 (see page 44). Indeed although nearly the full amount of additional money authorized by Parliament was issued in bonds, only some £30,000 of the £42,000 needed to complete the canal was received in cash – the balance representing consolidated unpaid arrears of interest on the original shares.

Although in August 1792 it had been hoped that 'with proper exertion' the canal would be fully opened by Christmas, this was far from being the case. On 3 September 1792 William Jessop had written to Lord Sheffield that 'the

BASINGSTOKE Canal Navigation,

April 21st 1794

WANTED IMMEDIATELY ABOUT FIFTY GOOD HANDS TO BE EMPLOYED ON VARIOUS PARTS OF THE LINE

GOOD wages will be given by applying to *Mr. GEORGE — SMITH*, The resident Engineer at *BASINGSTOKE*. *Mr. WEBB.* (*Mr. PINKERTONS*, Agent) at *ODIHAM*, or Any of the Agents on the Line.

By order of the Company.

CHARLES BEST.

An example of the canal's early mismanagement. George Stubbs's exuberance in advertising for fifty extra men 'immediately', more than four months before the canal was open, when it already had an establishment of thirty-three men, was ill-conceived. When, two years later, the company could not afford to pay its servants, Stubbs admitted that he was not the most competent of managers

Basingstoke Canal will soonest be likely to discharge its agents.' However, a notice in the *Reading Mercury* of 29 July 1793 of a house for sale in Basingstoke with an adjoining malt-house, referred to its 'peculiarly advantageous' position as the premises 'almost adjoin the intended wharf', which seems to imply that work on building the wharf had hardly begun. Progress had been delayed for want of finance and as a result it was not until the end of 1793 that all the locks were completed and tolls were being collected on traffic up to Odiham whose 3 acre wharf was let to James Hollis in December at £20 p.a. on a twenty-one-year repairing lease.[48]

A further delay arose when on a Wednesday or Thursday May night in 1794 the banks near Ash Bottom were maliciously damaged. The seriousness of the breach was such that an immediate reward of £50 was offered for information leading to the conviction of the offenders.[49] Consequently on 2 June the general meeting had no option but to agree that 'no person could be permitted or suffered to navigate any boat, barge or other vessel upon any part of the canal until after 24 June and that after that date no boat should be

The first lock on the Basingstoke Canal, at Woodham, 1967. The lock remained unworkable from 1968 until its repair in 1987

loaded to draw over 2 ft 4 in. until further notice'. It was also resolved that those intending to use the canal should attend at Basingstoke on the mornings of 7 or 21 June to receive instructions, register their boats and sign 'the rules, orders and instructions, without which none can be permitted to pass'.

It was customary for the opening of a waterway to be attended with great ceremony and considerable jollification. Barges would be decked out for the occasion, bands hired, church bells rung and a handsome dinner provided. For some reason or other none of these happenings appears to have occurred when the line of navigation to Basingstoke became navigable throughout. Certainly neither *The Times* nor the local newspapers alluded to the occasion. It certainly appears strange that there were no official celebrations held to mark its completion. One reason was possibly the fact that the proprietors with the largest number of shares – the Earl of Dartmouth, the Earl of Portsmouth and Lord Rivers – were not interested in taking part; another that there was no money available to meet the cost. The bare facts remain. In August 1794 announcements appeared in the press[50] that the canal would be opened on 4 September 1794, that barges would leave Basingstoke Wharf

The lock-house by lock III at Woodham, *c.* 1885. The barge-horses were stabled in the wooden building

Groom and barge-horse at Woodham, 1915

Horse-drawn barge at Byfleet, *c.* 1930. Horse towage continued until 1949

every Thursday morning at eight o'clock and from Messrs Sills & Sons, Hambro' Wharf, Upper Thames Street, every Thursday with the day's tide. Goods for dispatch would also be received at the wharves at Horsell, Pirbright, Frimley, Ash, Farnham Road (Aldershot), Crookham, Winchfield, Odiham, and Basing. The cost of dispatch in the company's barges was to be 12*s* per ton for the whole line, including tolls.

The total cost of building the Basingstoke Canal, excluding interest payments, was £153,462, or over £60,000 more than the original estimate made in 1777. A number of items such as wharves, warehouses and lock-houses had not been included in this estimate however, but even when £9,136 is deducted on this account, it will be seen that the canal had cost 50 per cent more than had originally been envisaged. The chief cause was inflation. Wages had risen steeply, especially for carpenters and seasonal employment as a result of labour shortages and price increases due to the American war. There had also been various contingent costs on account of the weather, malicious damage, the collapse of the banks at Greywell, etc. However, although the cost of land had not been excessive, 23 of the 37 miles being heaths or commons, it was a canal of some magnitude, spanned by no less than 68 bridges and possessing 29 locks, 5 lock-houses, 4 wharves, 3 warehouses and 2 tunnels.

Basingstoke Canal Navigation

NOTICE being given that THIS CANAL is intended to be opened on the 4th September, 1794; the Public are informed,

THAT Barges will begin to be Navigated on that day, from Basingstoke Wharf to the Hambro' Wharf, Upper Thames Street, London, and from London to Basingstoke, every Thursday.

The Barges from Basingstoke, will leave the Wharf every Thursday Morning at eight o'clock.

The Barges from London, will leave the Hambro' Wharf every Thursday, with the day's tide.

The Wharfs for receiving Goods on the Line of the Canal will be

At BASINGSTOKE,	FARNHAM ROAD,
BASING,	ASH,
ODIHAM,	FRIMLEY,
WINCHFIELD,	PURBRIGHT, AND
CROOKHAM,	HORSELL.

AND

At MESSRS SILLS and SONS, the HAMBRO' WHARF, UPPER THAMES STREET, LONDON.

The Rates for the Freight and Tonnage of all goods, &c. carried in those Barges, will be as follows, viz.

		For every Ton.				*For every half Ton.*				*For every quarter of a Ton and under.*		
		£.	*s.*	*d.*		*£*	*s.*	*d.*		*£.*	*s.*	*d.*
From BASINGSTOKE,	TO LONDON, —	0	12	0	-	0	6	0	-	0	3	0
From LONDON,	TO HORSEL, —	0	4	8	-	0	2	4	-	0	1	2
	TO PURBRIGHT, —	0	5	8	-	0	2	10	-	0	1	5
	TO FRIMLEY, —	0	6	8	-	0	3	4	-	0	1	8
	TO ASH — —	0	7	0	-	0	3	6	-	0	1	9
	TO FARNHAM ROAD,	0	7	8	-	0	3	10	-	0	1	11
	TO CROOKHAM —	0	8	8	-	0	4	4	-	0	2	2
	TO WINCHFIELD, —	0	9	8	-	0	4	10	-	0	2	5
	TO ODIHAM, —	0	10	0	-	0	5	0	-	0	2	6
	TO BASING, —	0	11	8	-	0	5	10	-	0	2	11
	TO BASINGSTOKE, —	0	12	0	-	0	6	0	-	0	3	0

EXCLUSIVE OF WHARFAGE, WEIGHING, LANDING, HOUSING, &c.

But the Owners of those Barges hereby declare, that they will not receive or be answerable or accountable for any Goods sent to be navigated by their Barges, unless the Numbers, Quantities, Qualities, Particulars, and Gross Weights, are marked thereon, or delivered therewith in Writing.

Nor will they be answerable or accountable for any Loss or Damage by Fire, Leakage, or other inevitable Accident in the Tide's Way, or otherwise.

Letters addressed to any of the Wharfingers will be punctually attended to.

N. B. The Wharfage to be paid before any Goods taken on board.

Notice of the opening of the Basingstoke Canal, 4 September 1794

The completion of the Basingstoke Canal heralded a new era in canal building. It was the first of the so-called agricultural waterways. Hitherto canals had been developed in the growing industrial areas around Birmingham and Leeds. Now the wealthier landowners of Surrey and Hampshire had built a canal: not to carry raw materials or manufactured products between the ports, factories and large towns, but to open up the countryside to enable the latest agricultural techniques to be used on the underdeveloped land – land which Defoe described as 'the great black desert, called Bagshot Heath'.[51]

The Revd Shaw, a keen agriculturalist, listed the benefits likely to accrue from the waterway:

> This being in the vicinity of many corn mills, and communicating with the most woody part of the county – and one of the best in England for fine timber – will be a great advantage. The mutual carriage of goods to and from the capital will be of great importance, and the west country manufacturers will find from hence an easy and cheap conveyance. An object of still greater importance is the likelihood of this canal being the means of promoting the cultivation of the extensive barren grounds before-mentioned through a great part of which it must necessarily pass, after having been first conducted through a country full of chalk, from whence the manure is now carried in large quantities, at the expense of one shilling per waggon load per mile; whereas by the canal it will cost but one penny a ton for the same distance; and the boats will return laden with peat and peat-ashes (the last are esteemed an excellent manure for saint-foin, clover etc) to the mutual benefit of cultivation, and the emolument of the proprietors.[52]

So much for expectations.

CHAPTER FOUR

TRADE DURING THE NAPOLEONIC WARS (1794–1815)

Disaster at Greywell (1794) – the Stubbs affair – financial crisis – interest on bonds suspended (1796) – the energetic Dr Bland, celebrated accoucheur – gift of plate – offices of the Basingstoke Canal Company – company's by-laws – cost of maintenance – threat of invasion – offer to help Government (1803) – state of trade – the company stops carrying (1804) – competition between barge and waggon – dividend on bonds resumed (1808) – purchase of Odiham Wharves (1813).

The canal company had no sooner begun carrying the 'goods, wares and merchandise of the public to and from London and Basingstoke with as much care diligence and attention, and with as much exactness and regularity as the nature of such extensive concerns and the situation of the canal would admit', then it was struck by disaster. Within six weeks of its opening the southern bank of the canal near the west end of Greywell Tunnel at Gruel Hill collapsed and 'this vast mass of clay' blocked the canal completely for a distance of nearly a hundred yards.* Although every effort was made to remedy the disaster, a second slip occurred and all trade was halted beyond the tunnel and for a time it was feared that a general stoppage might result. The only known copy of the report of the emergency meeting held in London in October 1794 is incomplete,[53] but traffic to Basingstoke was almost certainly stopped for the rest of the year and one presumes that a carrier service operated between Greywell and Basingstoke. Not until the following summer would it appear that the canal could be fully reopened. There had also been other difficulties. Alladay, the Weybridge lock-keeper, reported in December that the ice was so thick on the canal that the boats could not work through.[54]

During 1795 – the first year of uninterrupted operation – some 13,500 tons were carried, but in 1796 the very small demand for ships' timbers and the consequence of corn fetching a higher price at Winchester and the neighbouring markets than in London, reduced the tonnage carried by a third

* A similar blockage was narrowly averted in 1804 when a collapse reduced the channel to half its width, 'having laid the piles and planks with which it was attempted to be restrained almost flat'.

One SHARE £.10

RECEIVED 11 Nov 1796, of Mr Chas Langton the Sum of Three pounds being the Deposit on One Share of his Subscription for building Barges to be navigated on the BASINGSTOKE CANAL.

£.3 Thos Smith

Charles Langton's receipt for the deposit on one £10 share for building barges for the Basingstoke Canal, 1796. The company built five barges, which operated at a loss and were finally sold in 1804

and almost halved the toll receipts. Resort was had to local advertising and an announcement on 16 May rather surprisingly stated that 'the quantity of goods carrying on the canal far exceed what were expected at their early period and the bargemasters and others inclined to engage in the navigation are informed that the most flattering prospect is open to them of coming in to an immediate and considerable trade, and if such persons have not barges and do not choose to build, they may be accommodated with barges'.[55]

However, by the autumn of 1796 the company was on the verge of bankruptcy. Payments of interest on the bonds had to be suspended and if it had not been for the firm direction of the chairman, the company would have been forced into liquidation. Until this time the committee and shareholders had left the running of the company's affairs to George Stubbs. Stubbs was a Westminster solicitor who had also acted as attorney to the Wey Navigation proprietors since 1776. In the early days of the canal's history he had been responsible for negotiating agreements between the proprietors of the Wey and the promotors of the Basingstoke Canal. There is little doubt that he had played a leading role in getting the scheme off the ground and his name appeared in the Act as a commissioner appointed to settle differences between the canal proprietors and landowners regarding compensation.[56] The fact that his name had appeared on the prospectus published in June 1787 and that he took the chair at at least one general meeting (August 1792) when the practice was to choose different chairmen at each meeting, indicates that he had been one of, if not the principal, personality behind the company.

Indeed Stubbs's experience of the Wey Navigation had made him the obvious choice to be manager and superintendent of the canal company. He certainly made himself well-nigh indispensable. It had been Stubbs who 'had taken more pains than any other member and had acquired more knowledge of the business of the company'; Stubbs who had drawn up the petition for the second Act and had had sole management of the Bill while it passed through Parliament, and Stubbs who had been for several years chairman of the 'Committee of Accounts'.[57]

In August 1794 he had been appointed superintendent and from April 1795 was 'almost wholly' engaged in the company's business at Basingstoke. The problems were legion. The company had no offices, the clerk and the engineer both resided in Basingstoke and except for a small storehouse at Crookham, there was nowhere to store timber and materials to deal with any emergency between Basingstoke and the Wey Navigation. At the meeting in London on the last day of August 1795, Stubbs reported that the company needed full-time attention which he could not provide without making arrangements for his professional business in London and providing 'those comfortable conveniences in the country which his family have been accustomed to'. He therefore proposed to resign on 14 October unless he was offered a three-year contract. Faced with this ultimatum the proprietors could only accept his terms and he was granted the substantial salary of £600 p.a., which was to include any travelling expenses.[58] In due course he rented a house at Crondall.

The company had five barges operating between Basingstoke and London. Apparently these barges had no living accommodation and Stubbs suggested that the proprietors should consider building houses for the bargemen and their families as close to the line as possible to keep them from wandering, or leaving the work as was too frequently the case, to visit their families at 20 or 30 miles distance.[58] Stubbs pointed out that there were no houses on the line and that by attracting persons needed for the navigation to the neighbourhood, their children might by degrees be initiated in the business.

It had been Stubbs's recommendation to the proprietors in the spring of 1797 that they should apply once again to Parliament for further capital, as in his opinion that was the only means by which the business could be continued. Only at this eleventh hour had the committee and shareholders realized their predicament, for it was now clear that the company had no money, was more than doubly pledged and unable to borrow more except from among themselves. As early as August 1796 Stubbs had hinted of his intention to resign unless more capital was raised. 'It is with great concern that Mr Stubbs is compelled in every report to remind the company of the want of money to carry on and manage their affairs; and it is more painful to him now to declare that without the necessary supplies, he can no longer continue in the situation they have been pleased to place him.' It is perhaps significant that Stubbs was prepared to make one admittance – 'he is sensible that much more might have been done, and that his abilities were not so competent to the management of such an undertaking as might have been found in others.'[59] Nevertheless, he was anxious to point out that should the company decide to 'desert the whole undertaking, they must not impute their neglect or misfortune to him'. Stubbs, however, met his match in the person of the newly elected chairman, Robert Bland, and at the general meeting in February 1797 Stubbs was overruled when 'contrary to his opinion and formal protest', the company resolved to let all its barges. Claiming that only one sixth of the voting strength was then present, he circularized the proprietors of his intention to resign 'the sole superintendency' but to continue to attend general meetings or committees and requested their proxies so that they could trust their vote to his discretion.[60]

The reason for the company's plight was not difficult to discover. Matters had been left for far too long in the hands of Stubbs. Although the committee and shareholders had no one but themselves to blame for this state of affairs, as they had approved the superintendent's previous proposals, they indignantly pointed out that he had taken advantage of their complete confidence in him and that they had had no reason to suppose that he was not acting in their interest and contrary to the Act of Parliament which he had solicited and obtained. Indeed it is interesting to note that the minutes of the meeting on 29 February 1796 stated that the proprietors 'are abundantly satisfied with the zeal and ability with which he [Stubbs] has continued to conduct the affairs of the company: and although from the exhausted state of the finances, they have been obliged to defer carrying into execution for the present part of his plans, they are nevertheless convinced that they were highly eligible and proper, and hope in a more favourable season to adopt them'.

Briefly, the situation which had arisen was that after the company had obtained authority in 1793 to borrow a further £60,000 to complete the canal, Stubbs had continued to pay interest on shares issued under the original Act, which the second Act had stipulated should not be done until the objects set out under this Act had been carried out. However, the committee had indeed approved and the shareholders had accepted the superintendent's plan for issuing bonds for which cash was paid only in part, the balance being met by the issue of receipts for arrears of interest. Consequently the company, while obtaining about £31,000 in cash, also gave these subscribers interest-bearing bonds to the value of £24,000 and in doing so not only created a further annual charge of £1,200 but effectively prevented the *raison d'être* for the second Act, the completion of the canal, since the Act allowed no more than £60,000 to be raised, and at least £41,700 was needed for this purpose in cash.

The true position was hidden from the company by Stubbs, by whom, said Dr Bland, 'we have frequently been told that the interest of the bonds was paid out of the tonnage tolls'. But, the chairman pointed out, that was a fallacy, since during the whole time the canal had been managed by the superintendent it had been found that there had been no profit, 'only a deficit of several hundred pounds per annum'.[61] Ruin had in fact only been averted by the specialty creditors agreeing to suspension of interest on the bonds notwithstanding vehement opposition from Stubbs and one or two others. In fact when Dr Bland had chaired a meeting in February 1796 to consider whether or not the canal could be said to be completed according to the Act, so that interest could cease being paid on the shares, the resolution was deferred and a counter-motion carried that the 'proposition should not be considered'. The point at issue was whether the canal could be said to be completed when the Turgis Green branch had not been begun. Counsel's opinion was sought, which merely confirmed that without the consent of the majority of the proprietors present at a general meeting, interest would have to continue to be paid.

It was difficult to prove any allegation against Stubbs or indeed to discover the true facts – and debts – since the superintendent had taken all the books with him when he was dismissed. Stubbs defended his actions, moreover, on the grounds that the proprietors had no right to suspend interest payments on the bonds and that if the company could not pay the interest, it should go into liquidation so that the creditors could claim their debts.

When Dr Bland put forward a scheme to salvage the company's finances – the redemption of bonds by issuing new bonds for only the amount of the cash actually paid plus five years' interest – Stubbs distributed a printed circular (18 August 1800) strongly dissenting from the plan, justifying his conduct and suggesting that facts were being suppressed by the committee. The shareholders thereupon unanimously passed a resolution:

> That this meeting are of the opinion that no industry hath been used to pervert, misrepresent or suppress facts, but that the chairman and committee have at all times been ready to give every information in their power, and are entitled to the thanks of the company. Mr Stubbs having stated that it does not appear that the affairs of the company have been progressively mending within the past three years, this meeting begs leave to refer the proprietors to the fact that trade has been regularly increasing during that time and that the expenses have been diminished.[62]

The exact number of bonds issued, said the chairman, 'for want of having the books in which the accounts were kept, in our possession, cannot be known'.* In addition, Stubbs had contracted debts of about £7,500 with the company's banker and tradesmen in order to pay interest on the bonds. It took five years to clear these debts and to put the company's finances on a proper footing.

In the early days of the canal's history the affairs of the company were, as has been described, very much in the hands of George Stubbs. However, with his dismissal in 1797 the task of running it was undertaken by the chairman, Dr Bland, with the assistance of the clerk and a management committee composed of twenty proprietors each holding at least three shares. Since, however, the committee mainly consisted of country gentlemen and professional men without the time or inclination to occupy themselves with the details of managing a waterway, as 'the majority of the members were stationed at some distance from the canal', the company's fortunes depended upon the chairman. In this respect they were more than fortunate.

Robert Bland was the son of a Norfolk attorney and at the time of his entry into the canal's affairs, when in his fifties, was well known in London as a distinguished gynaecologist and as a contributor to various medical and philosophical journals.[63] Little is known of his experiences before he qualified as a doctor at the age of thirty-eight, except that he studied at St Andrews University and trained at various London hospitals. He developed an extensive practice in London as an accoucheur, and was one of those responsible for founding the Westminster General Dispensary in 1774. About this time he married, and the first of his four children was born in London in 1779. He was made a licentiate of the Royal College of Physicians in 1786 and acquired a high enough reputation to be commissioned to write articles on midwifery for Rees' Cyclopaedia. A fellow of the Society of Antiquaries, there appeared in 1794 his *Observations on Human and Comparative Parturition* and in 1814, his *Proverbs chiefly taken from the Adagia of Erasmus, with Explanations and further illustrated by corresponding examples from the Spanish, Italian, French and English languages. The Monthly Review* commented that this work is adapted to supply the reader with reflexions, the converser with quotations, the writer with metaphors and the moralist with rules of life.

Nicholas Carlisle said: 'he was skilful in his profession, and of extensive experience but eccentric in his manner'. Biographies, however, make no reference to his close association for over twenty years with the canal company, nor do they give any clear indication as to why he should have devoted so much of his time to trying to put the company on its feet. The fact that the Earl of Dartmouth was president of the London Dispensary may

* The estimated figure was £55,000. Some of the papers were returned to the company by Stubbs's executors in 1808. (*BCR*, 9 February 1809.)

At a General Meeting of the Company of Proprietors of the BASINGSTOKE CANAL NAVIGATION, *held by Adjournment at the Crown and Anchor Tavern, in the Strand, London,*

ON THURSDAY, DECEMBER 8, 1796,

DR. ROBERT BLAND IN THE CHAIR.

[*Extracts from the Minutes of the said Meeting.*]

MR. STUBBS presented the following progressive statement of the tonnage on the Canal, to Michaelmas 1796:

					TONS.
In the year 1791 there were carried on the Canal between the River Wey and Horsell					28
In the year 1792 there were carried on the Canal between the River Wey and Purbright					173
In the year 1793 there were carried on the Canal between the River Wey and Odiham					857
In the year 1794 there were carried on the Canal between the River Wey and Basingstoke, viz.					
In various barges				4366	
In the Company's barges				1431	
					5797
In the year 1795 there were carried on the Canal between the River Wey and Basingstoke, viz.					
	TONS			TONS	
To Lady-day, in the Company's barges	511		In other barges	697	
To Midsummer, ditto	1349		Ditto	2325	
To Michaelmas, ditto	1552		Ditto	2159	
To Christmas, ditto	1364		Ditto	1764	
	4776			6943	
			Making together		11,719
In the year 1796 there were carried on the Canal between the River Wey and Basingstoke, viz.					
To Lady-day, in the Company's barges	1334		In other barges	1863	
To Midsummer, ditto	1662		Ditto	2630	
To Michaelmas, ditto	1783		Ditto	2152	
	4779			6645	
Making together, for the three quarters,					11,424
Total carried on the Canal since it was cut					29,998

Minutes of the meeting held in London on 8 December 1796, stating the tonnage carried on the canal since 1791 have led to his initial interest in the canal and it is not unreasonable to assume that as a change from the complications of childbirth, he became fascinated by the problems of management which added breadth to his interests, as well as providing him with what turned out to be a not unwelcome supplementary income. After his death Sotheby's sold his extensive library 'consisting of a very good collection of books in medicine, chemistry, arts; likewise Latin facetiae, classics, etc'. Perusal of the auction catalogue of this four-day sale throws no light on Bland's interest in the waterway. He did, however, possess a copy of *L'indécence aux Hommes d'accoucher les Femmes.*

It took some time to reorganize the company's affairs, but within twelve months of Stubbs's departure a loss of £1,200 had been turned into a profit of £614. 'By giving up and letting the barges, by saving the salary of the superintendent, dismissing one clerk, reducing the salaries of others', savings of over £1,000 p.a. had been made. Trade continued to develop and in the twelve months ending on 25 March 1801 income exceeded expenditure by £2,038. No provision was made, however, for depreciation.

The minutes of the management committee's quarterly meetings relate the efforts which were made to improve the running of the concern. Problems had arisen due to the indifferent work of some of the sub-contractors which had to be remedied. The lack of draught above Greywell materially impeded the navigation and threatened the structure of the tunnel 'in consequence of the force necessarily used by the men against the sides of the arch in pushing their barges through when short of water'.[64] Consequently in 1798 a lock (lock XXX) had to be built 200 yd from the east end of the tunnel to keep up the water-level. The committee reported in the autumn of 1799 that, notwithstanding the continued and heavy rains, which had done so much mischief in many parts of the kingdom, no material accident had happened on any part of the canal, and that whereas the Thames & Severn Canal and Kennet Navigation barges could not use the upper part of the Thames for nearly six weeks, the company's barges had suffered no interruption, and on the contrary, 'favoured by the great abundance of water the rain has afforded, they have been able to carry larger loads than at any former period'. This optimism was premature. At the January meeting it was reported that four additional barges had had to be employed to clear away the goods that had accumulated during the floods and frost. Later the same year a drought was reported that had 'exceeded in intenseness and duration any drought that has occurred for many years past'. Barges were restricted to half loads, but as the water-levels were equally low on the upper Thames, the committee reported that there was no fear of trade being lost to the Kennet Navigation.

The committee's endeavours were not only directed towards increasing trade, but also to reducing maintenance costs, 'which had been conducted on

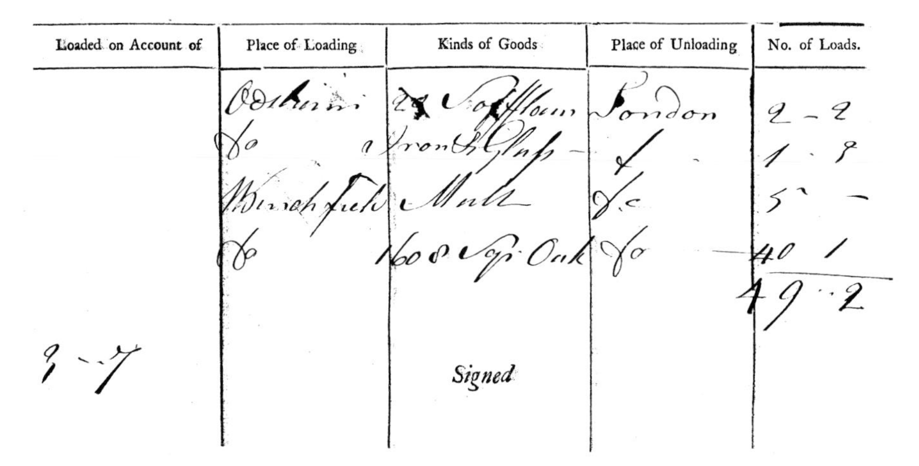

RIVER WEY, SURREY.

No. Dat 9 May 1797

Loaded on Account of	Place of Loading	Kinds of Goods	Place of Unloading	No. of Loads.
	Odiham	2? Sacks Flour	London	2 - 2
	Do	Iron & Glass	Do	1 . 8
	Winchfield	Malt	Do	5 -
	Do	1608 sqr Oak	Do	40 1
				49 . 2
9 - 7		Signed		

Wey Navigation way-bill of 1797 for a varied cargo of flour, iron, glass, malt and 1608 sq ft of oak timber, loaded at Odiham and Winchfield and destined for London

a scale much too large for the income of the company'. Their difficulties were both numerous and 'not easily surmounted, as they had not only to break off habits which from inattention or indulgence had become inveterate, and thence to encounter the reluctance and perhaps ill-will of the persons whose emoluments were to be abridged', but to acquire considerable knowledge about the working of the canal. Much of this knowledge had been acquired by Dr Bland who had apparently been making twice-weekly visits to the Hambro' Wharf in Thames Street to ascertain how the men were employed. While goodwill and economic sanction seldom make good partners, it is some measure of his success that the proprietors, on being informed that the office of chairman was attended with very considerable trouble, presented Dr Bland with a gift of plate worth 100 guineas – 'a small token of acknowledgment of their sense of the benefit the Company have derived from his zealous and unremitting exertions in promoting the interest of their concerns'. Shortly afterwards it was agreed to pay the chairman 100 guineas a year, but later this was rescinded on the recommendation of the committee that Dr Bland should receive 200 guineas 'for his care and attention to the business of the Company during 1802', and that he should continue to receive this sum 'so long as he shall continue to manage the business of the company'.

The secretarial work devolved upon the clerk, Charles Best, a solicitor and Town Clerk of Basingstoke, who received a salary of £60 p.a. He had a deputy, William Harrison, who looked after the office in London. Harrison resigned in 1803, 'being obliged to go and reside in the country'. The committee, after interviewing several candidates, appointed Peter Jolit to fill the vacancy. Described as 'a young man about twenty-five years, married, of respectable connections and who can give ample security for the trust to be reposed in him', he was recommended to the shareholders as 'competent to fill the office with credit to himself and advantages to the company'.

The company had initially used Stubbs's office in Suffolk Street and later in Queen Street, Westminster, but after the rift in 1797 it had rather hurriedly moved into Charles Street, off St James's Square. The latter building, it appears, was pretty miserable, for when the lease expired at the end of 1803 it was said the house was in a 'ruinous state and intended to be taken down'. The canal office was then transferred to Hambro' Wharf by Three Crane Stairs at the bottom of Queen Street, Cheapside. This move had the advantage of enabling Jolit to supervise the dispatch and receipt of goods and, by knowing the state of the packages, resist 'improper claims for damages or

The wharves above Southwark Bridge, 1825. Barges were loaded at Hambro' Wharf and fly-boats, travelling by day and night, left Kennet Wharf twice a week for Odiham and Basingstoke. A plan showing the location of the wharves is on page 91. Hambro' Wharf is located to the immediate left of the bridge

losses which the company have not infrequently been hitherto obliged to admit and pay'. This arrangement did not last long, however. After the company had decided to let its own barges, the office on Hambro' Wharf was closed in 1806 and Jolit sacked. George Adams, the wharfinger at Basingstoke who had been responsible for 'the superintendence' of the company's barges and the management of the trade, was made responsible for the tonnage account hitherto kept in London; his salary was consequently increased from £90 p.a. to £100 p.a. At the same time a young man, J.R. Birnie, was recruited to act as deputy clerk at £30 p.a. instead of £80 p.a. which had been Jolit's salary. By 1808, however, the London office had been re-established on the corner of Spur (now Panton) Street and Leicester Square – in part of Dr Bland's house* – where it remained until his death.

The general meetings of the company were initially held every quarter. All, except the June meetings which took place at the Crown Inn at Basingstoke, were held at the Crown & Anchor Tavern in the Strand. The August meetings were, however, thinly attended and after 1802 the annual number was reduced to three.

The company's by-laws stipulated the rules of the navigation and highlighted some of the waterway's limitations. For instance the canal could not be navigated at night, being closed in summer between 9.00 p.m. and 4.00 a.m. and for twelve hours a day from 6.00 p.m. in winter. Furthermore, boats could only enter Greywell Tunnel if bound for Basingstoke between 4.00 a.m. and 5.00 a.m., noon and 1.00 p.m., and 8.00 p.m. and 9.00 p.m., and if bound for the Thames between 8.00 a.m. and 9.00 a.m., 4.00 p.m. and 5.00 p.m. and midnight and 1.00 a.m.. (This latter hour would appear to contradict the rule about no night navigation.) To avoid damage to the bed of the canal, poles used for navigating the Thames, which were shod with iron, could not be used upon pain of a 40*s* fine. Up boats had to give precedence to down, unloaded to loaded. Timber could not be floated, guns and fishnets were prohibited. Boatmen navigating rudder foremost could be fined five shillings. Wilful damage to the canal or diverting a feeder stream was judged a felony. The severest forfeit of £10 plus the cost of repair was rightly reserved for damaging a lock gate, while bathing or throwing a dead dog or cat into the canal carried a penalty of not less than 5*s* but not more than 40*s*.

The expense of maintaining the canal was inevitably considerable, as besides the thirty locks and seventy-two bridges, there was a prodigious number of culverts beneath the canal to be kept in repair. Severe weather gravely hampered traffic and more often than not additional labourers had to

* Now demolished, the site was later occupied by Fanum House, the Automobile Association's head office until 1971. By a strange coincidence the author wrote this book while working at Fanum house.

be employed in winter. Three 'dirt' barges were held in readiness for any emergency. A dredging and weed apparatus was bought to speed up the toilsome manual process of keeping the waterway clear during the summer.

Every August the management took out the company's pleasure barge for the annual survey. Repairs which had been carried out were inspected, the condition of banks, bridges, lock gates and aqueducts noted. The reports were often gloomy: 'While this account is printing, by the continuance of the wind and rain, the brick facing of the head of Peat Pond, at Pirbright, was washed down to the extent of about thirty yards; fortunately the bank was prevented giving way, otherwise the damage would have been very great; checked as it has been, it will probably be repaired for about three score or four score pounds.'[65] A few years later the chairman reported that 'the proprietors will not be surprised to hear that the banks of the canal have suffered very materially from the late inundations. The repair of damages incurred will add very materially to the expenses'. In the autumn of 1812 the collapse of 24 ft of the crown of the arch of the tunnel and the discovery of a serious fissure in the embankment over Ash valley occasioned no small alarm. The cost of meeting such contingencies as well as general maintenance represented a substantial proportion of the canal's revenue; not only did major repairs involve expense, but trade was stopped until they were completed.

The following winter ice blocked the canal twice before Christmas; 'since then it has been completely frozen over to the present time [mid-February], and as the fall of snow has been unexampled for quantity, there has been not only an entire cessation of trade, but we have been obliged to employ additional labourers, above those that are kept in constant pay, to clear the drains, unload the banks, open the mouths of the culverts and lower the weirs'. The company was in some respects fortunate when towards the end of 1814 'the rough and boisterous weather', which had occasioned so much destruction among the shipping, damaged the heads of two of its ponds. 'An immense quantity of water rushed out, with such an impetuosity, as would probably have overturned any moderate building that had been near, but it was all scattered, and lost in a barren heath, so that the expense incurred will be very little more than what will be required to repair the pond heads.'

The swing-bridges, 'the most perishable of the works', had to be replaced every fifteen or twenty years; for failure to repair the footbridge at Mapledurwell in 1818 the company was indicted at the Quarter Sessions but obtained a stay of execution while it was rebuilt.

Of the many complicated and expensive works, it was the south bank at the western end of the tunnel which was the cause of greatest concern. After its initial collapse in 1794 and again in 1804, it was reported in 1807 that although many thousands of tons of clay had been removed annually, 'it was still moving forwards, threatening to pull down with it part of the tunnel'.[66] Five years later Dr Bland observed that the slip west of the tunnel was continuing its encroachments. 'Directions were now given to drive piles, a great depth, viz nine feet into the ground, and then to face them with strong oak planks'. But to little avail for in 1824 it was again necessary to remove the clay and replace it with chalk and gravelly loam, making drains in various directions, covering, turfing and planting where practicable.

The Peace of Amiens in 1802 had brought mixed hopes to the proprietors, who could not refrain from commenting in their report that 'the return of peace, by giving security to the coasting trade, may induce many of the customers to the canal to prefer that mode of sending their goods, as being much cheaper than by the Canal'. To counterbalance that advantage it was planned to make carriage by canal more regular, expeditious and certain than by sea. It was therefore perhaps more than unfortunate that on 22 March that year the *Baxter* should have sunk after encountering a sudden and violent storm in the Thames about Nine Elms.* The company's barge had been loaded with groceries and merchandise and while the committee expressed their concern at so heavy a disaster, they were quick to take pains to assure the proprietors that the company was not liable to make good the attending loss; 'the bill given to the shippers on receiving the goods, declaring they will not be answerable for any loss occasioned by fire, leakage or any unavoidable accident'.

However, the underlying problem still remained the shortage of water, which not only added to the risks of pilferage but had caused stoppages and the heavy expense of providing vessels to lighten the barges. Consequently the committee directed a surveyor, Henry Goolding, to examine those places most likely to produce an additional supply and also to discover, if he could, any leakages that might have escaped the vigilance of the company's surveyors. Later that year the committee was to report that while the amount of cargo carried had not decreased, receipts had, and that this was one of the consequences of the peace, which by diminishing the risk of the coastal passage had caused a considerable quantity of the goods consigned to the Isle of Wight, Jersey, Guernsey, etc, which used to go by the canal to Basingstoke, and thence by waggon to Southampton, to be now sent directly from the Thames in vessels trading to those islands.

Local trade had, however, increased, for it was reported that the additional quantity of goods sent up the canal to Bagshot, Farnham and Odiham in some degree compensated for the loss of trade to Basingstoke. Nevertheless, although the total tonnage carried in 1802 was only 2,500 tons less than the

* Another barge belonging to Mr Jones, and three other vessels also foundered that day.

previous year, tolls were down by nearly 20 per cent and the profit for the year halved.

But the peace did not endure and in May 1803 the war had been resumed. This was fortunate from the company's point of view. 'During the short interval of peace enjoyed by the country, the Committee had the mortification of observing a manifest decline in the trade, particularly that which went the whole length of the canal to Basingstoke and consequently paid the largest tonnage.' However, with the return of hostilities, trade quickly improved. Since Napoleon was now bent on subjugating England, the threat of invasion loomed large. That autumn several shareholders in London approached the committee and suggested that following the example of other corporate bodies, the company should offer the government whatever assistance it could towards transporting baggage and stores up and down the canal. After the consent of the bargemasters working on the canal had been obtained, the committee wrote a letter, which the chairman and Mr Baker personally delivered to Lord Hobart, the Secretary for War and the Colonies, offering to make available ten Basingstoke Canal barges free of charge to transport stores from London to any part of the canal 'in the event of invasion or the appearance of the enemy on the coast'. Lord Hobart gratefully acknowledged the committee's kind offer which, as events proved, the government did not need to take up.

Although the merits of an inland water communication between London and Portsmouth were widely mooted, no steps were taken by the government to join either the Wey and Arun Navigations or the Basingstoke and Itchen Navigations. A survey was made for a broad canal from Croydon to Portsmouth in 1803 but it failed to attract support. However, as a means of both providing a defence against an invading army and of moving troops and stores along the stretch of coast most threatened by the enemy, work was begun in 1804 on the Royal Military Canal from Shorncliffe to Winchelsea.*

In the summer of 1804 the chairman reported that there had been a considerable fall of timber in the neighbourhood of Odiham and Winchfield which was being brought down to the wharves and that a new source of trade had been 'lately opened, with malt and flour from King's Clear [Kingsclere]. . . . The upward trade, on the other hand, consisting of coals, yarn and other articles, from the Baltic, with grocery etc, owing to the pressure of the times, has been some time declining. . . . The deficiency in the carriage of coal has probably been occasioned by the high price that article has for a long time borne.'

* See P.A.L. Vine, 1972, *The Royal Military Canal.*

By 1806 the committee felt able to trust that the accounts 'now laid before the proprietors will show them that their property is not in so hopeless a state, as it has appeared to them, and as they are accustomed to represent it to themselves'. In 1807 the company was anticipating an annual operating surplus of about £1,400 and was for the first time able to announce that it was now in possession of all the land covered by the canal that was purchasable. Robert Bland paid tribute to 'those gentlemen for the loan of £5,000 lent at a time when the company could neither have well done without such assistance nor could easily have found any other mode of obtaining it'. These gentlemen included the Earl of Dartmouth, Farmer Platt and the clerk, Charles Best.

In July 1807 the proprietors were invited for the first time to join the committee on their annual survey of the line and to travel on 'the company's barge'. After the general meeting, accommodation was arranged in Basingstoke prior to a two-day cruise to Weybridge, the second night being spent at Farnham. In 1808 payment of 2 per cent was resumed on the bonds to those specialty creditors who presented themselves at the offices of the company with their bonds ('so that they may be compared with the entries of the books of the company') on Tuesdays between ten and two o'clock.

Claims for compensation sometimes took many years to settle. The collapse of the brick heads of two ponds at Pirbright in January 1804 (see page 52) had damaged two of Mr Halsey's peat-grounds. Dr Bland reported in October 1808 that he had had 'frequent' meetings with Henry Halsey of Henley Park, Farnham, 'but his demands were so exhorbitant that he could not settle it with him'. Not until a year after Halsey's death was composition for the damages (£200) agreed with his executors.

During the first ten years of full operation tolls averaged £3,000 p.a. representing the carriage of about 15,000 tons per year. The company had several additional sources of revenue: in 1796 Thomas Jefferys, one of the larger shareholders, had been granted a twenty-one-year lease to carry away chalk from any of the company's lands free of toll on payment of £1,000 and an annual rent of £300.* Income was also derived from the rent of Fleet Mills and various strips of land, from the loading service provided at Basingstoke Wharf, from the sales of bavins, furze and rushes and, until 1804, from its carrying business – about £200 p.a. all told. The company's own barges had, however, run at a loss. At certain times of the year, particularly when the canal was suffering one of its dry spells, the bargemen found it uneconomic to trade, so that the company had to run more barges than it would otherwise,

* This arrangement ceased in 1806 when Mr Jefferys, having 'moved into Gloucestershire', gave up the lease on payment of over £1000 compensation.

'and it will reasonably be supposed, that if men navigating their own barges scarcely earn a subsistance, the company, who must pay for everything at a higher price than those men do, besides the impositions they are liable to, must be great losers by this part of their trade'. During the year ending on 25 March 1804, losses amounted to over £1,000 from this business; consequently in the autumn the company ceased trading, 'having found the expense of navigating, which they could no way control, prove a sink, swallowing up a large portion of the tonnage. It made their accounts voluminous and difficult, obliged them to keep additional servants, upon whom they were in a degree dependent. The experiment has in every point succeeded.'

The surplus of income over expenditure was, however, never very substantial. Although it exceeded £2,000 in 1800–1, the average for the decade was little more than £1,000 p.a., representing earnings of less than 1 per cent on the original equity and ignoring the interest payable on the bond debt. This presented a very different picture to that painted by the promoters in 1787 when they estimated the revenue from tolls alone at over £7,500 p.a. But later the committee did point out that these calculations included trade on the collateral cut to Turgis Green and that the tonnage was assumed to go through the whole extent of the canal. These two items, they averred, accounted for a deficiency of about £3,000 p.a. Against this though, they claimed that the trade from London to Basingstoke was much more productive than had been anticipated and that, in addition, £300 p.a. was already received for the sale and carriage of chalk from the company's land. In 1797 they were still anticipating an income exceeding £6,000 p.a. within a few years – but this expectation was never fulfilled.

The third volume of Manning and Bray's *The History and Antiquities of the County of Surrey*, published in 1814, after referring to the unfortunate persons who 'ventured their money on the flattering and fallacious representations of the schemers who undertook it', also mentions that the building of the military college at Sandhurst soon after the turn of the century 'procured an increase of carriage' on the canal.*

There were still nasty surprises in store. In October 1811 the proprietors were advised that the Commissioners for the Property Tax had requested a

* Although the land and manor house at Sandhurst were bought by the Government in 1801, the building of the college took eleven years to complete. Apparently the contractor, Alexander Coupland, had done little more by 1807 than set up his own brickyard to produce many thousands of bricks to avoid the cost of transporting them from London. The building, however, was not begun until 1808 when, Coupland's bricks being found useless, a new supply costing £4,000 was ordered from London. The main block of the academy was completed in 1812 at a cost in the region of £350,000. It was during this latter period that the canal was of most use. (Hugh Thamas, *The Story of Sandhurst*, 1961, p. 37, *et seq*. and John Selby, 'Royal Military Academy, Sandhurst', *Country Life*, 26 June 1969.)

return of their profits. As the committee supposed that they were not liable to pay that tax due to the receipts being 'insufficient to pay the interest of their debt', they were saddened to learn that they had been assessed £1,130 for the year ending the previous April.

Competition was now increasing from the waggon trade. By 1810 the cost of carriage by canal between Basingstoke and the metropolis had risen to 18*s* per ton while that of land carriage had fallen to around 35*s* per ton. Although the company ceased to run its own barges in 1804, it nevertheless suffered losses through the failure of some of the traders: three ran into difficulties in 1813 (the company had to take their barges over and let them) and a fourth – 'a very worthy and respectable trader lies now speechless from a paralytic affection and whatever may be the event, it will be likely to lock up more than £600 for some time to come'. Bland's concluding remarks that 'it will be pleasing to the proprietors to know that these accidents occasion no other inconvenience to the company than that of retarding their progress a little in extinguishing the bonds', suggest a degree of naivety; the loss of traders could only reduce traffic and the tying-up of capital affect the company's earning capacity. Indeed, sixteen months later Bland reported that

> among our recent disasters, it will be proper to say, that by the death of two of our traders, and by the failure of two others, we are likely to be considerable losers, to the amount, it is probable, of £800 or £900. By temporizing and giving the parties time to part with their trade and stock, the trade was preserved, and transferred to more substantial adventurers and the debts were considerably reduced. By a different conduct, something more might perhaps have been obtained, but the credit of the business would have suffered and a great part of it have been lost.

The company husbanded its resources by sowing furze and planting timber and underwood along the banks of the canal. 'Much of these are growing, and in time, the whole of the slopes will be covered with herbage, which will not only keep up the banks, but yield a small annual profit.'[67] The fir plantations established at Basing, Nately and Greywell totalled over five acres. In 1822 Cracklow, the Wey agent, suggested to the joint owners that they should do the same 'in the manner judiciously adopted by Mr Birnie along the Basingstoke Canal'; and also that quick hedges should be substituted for expensive wood fencing which was being continually plundered by the poor for firing.

In 1805 a new lock-house was built at Frimley and in 1813 a shed for repairing barges was added; in 1806 it was decided to build a 'small house' for George Webb, the wharfinger at Odiham who was 'an old diligent and

The Great Wharf, Odiham, 1867. The carts appear to be loaded with sacks of coal. The sale particulars of the canal in 1869 referred to the 'well-arranged and substantially built wharfinger's residence, counting-house, brick and tile-built range of stores and warehouses, stabling, crane and shed'

faithful servant'; also planned was a house on the wharf at Basingstoke for the book-keeper who had served the company from the time the canal was first opened. Lack of finance made it difficult to provide all the necessary appurtenances. Not until 1812 was the 'long and much wanted' house built for the lock-keeper at Ash. As trade developed, coal-pens, kilns, saw-pits and cranes arose along the canal. Besides a number of warehouses and counting-houses which stood on all the big wharves, Mapledurwell boasted a picturesque timber and thatch storehouse, Crookham a timber-store and coal-shed while at Woodham a cow- and tackle-shed stood adjacent to the thatched stables. It was reported in 1813 that the wharves at Odiham, which were not owned by the company, had fallen into decay in recent years. Robert Bland commented after visiting them in July that they 'exhibited a face of distress and desolation and were in want of almost every kind of convenience . . . having been for many years in the hands of an indigent and very indolent man'. Consequently most of the traffic to Alton and beyond had been driven to the Farnham Road Wharf at Aldershot to which point it paid a much reduced toll. This wharf was capable of being made the second in point of advantage in the canal, so the company went ahead and bought both the Great Wharf and the Little Wharf at Odiham for around £1,000 and spent further substantial sums purchasing the crane and other 'erections', rebuilding the warehouse, establishing new coal-pens and building a house for the wharfinger. This work was completed in the autumn of 1815 and the purchase of bonds, temporarily suspended in 1813 on account of this expenditure, was resumed in 1817.

The war had only brought a limited increase in traffic but prices had soared; the first barges the company built cost £140; those in 1814, £320. On 18 June 1815 the conflict was finally won at Waterloo, but while other navigations had shown substantial profit as a result of the increased traffic created by the needs of war, the Basingstoke Canal had in the course of its first twenty years of operation shown that it could do little more than meet expenses and that while it was saddled with its enormous bond debt, it was unlikely to pay a dividend. That was, unless it could be more profitably linked with another line of navigation.

Little Wharf, Odiham, *c.* 1910, when pleasure boating was a popular recreation on the upper reaches of the canal

CHAPTER FIVE

THE ATTEMPT TO REACH THE CHANNEL (1790–1810)

The need to extend the Basingstoke Canal – John Rennie's report (1790) – plans to reach Andover and Winchester (1792–6), Turgis Green (1797–1800) and Bagshot (1801) – state of the Thames Navigation – cut proposed from Kingston to the Wey Navigation (1802) – Portsmouth, Southampton & London Junction Canal project (1807) – links proposed with the Itchen and Wey Navigations – £124,000 subscribed – reasons for failure.

It was never intended that the Basingstoke Canal should remain a navigation to serve only London and central Hampshire. That it failed to develop was the result of circumstances and certainly not for want of plans or imagination. For over thirty years proposals were made for linking the canal with a variety of towns, with the Thames, and with the English Channel. As early as 1783 plans were afoot for making Southampton the centre of the canal system of the south coast with through workings to the Thames and London in the east and to Bristol in the west. The Basingstoke committee stated that it was 'scarcely doubted' but that before the navigation to Basingstoke was completed, another company would continue it to the Bristol Channel on one side and 'into the British Channel by Southampton or Christchurch, with an arm to Salisbury' on the other.[68]

When it is realized that this 37 mile long canal served at this period only one town besides Basingstoke along the whole line, and that the population of Odiham was fewer than 1,500, and when it is remembered that both Aldershot and Woking at this time were only small villages with fewer than 500 inhabitants, it is readily evident that the only remunerative traffic likely to develop would be centred on Basingstoke as the distributing point for the market towns of Hampshire and Wiltshire. Only in time of war was the canal likely to be used as the principal means of transporting goods between London and Southampton. It will therefore be appreciated how small was the potential water traffic of this agricultural canal unless it could be expanded; and in this respect the company's expectations had very nearly been prematurely ruined by the actions of its superintendent, George Stubbs.

The passing of the Andover Canal Act in 1789 to link Southampton to Andover created speculation about an extension to Salisbury and Basingstoke. The Basingstoke proprietors, who attended a meeting at Andover in May, had to advise the promoters that their Act did not permit them to contribute towards the expense of the survey.[69] However, Alexander Baxter, the chairman of the management committee, together with John Granger, the agent of the Wey Navigation and other Basingstoke proprietors, went ahead and formed a committee to press forwards with the idea of extending the canal to Salisbury and, as Jessop was fully engaged, invited John Rennie to carry out a survey.

Before accepting, Rennie very politely wrote to 'Mr Jessop engineer Basingstoke' stating that as his first acquaintance with that part of the country originated in you, I should consider myself unpardonable in accepting anything in which you have been employed without you being apprised thereof' and concluding his letter 'Trusting that this information will go no further'.[70] Rennie, not yet thirty, was at the beginning of his career and had previously only been consulted regarding the Bishop's Stortford Canal, whereas Jessop, sixteen years his senior, was now the foremost canal engineer in Britain. Rennie had been overjoyed to receive the invitation and, on receiving Jessop's favourable reply, wrote to Baxter stating that he had just learnt that he owed the appointment principally to him. 'I therefore beg to return you my best thanks and to aprise [*sic*] you it has made the most lasting impression on my mind – the more so as my short acquaintance scarcely left me room to expect so great a mark of your favour'.

Rennie took the levels between Basingstoke and Polhampton and made an 'occular' survey to Andover towards the end of January. The problem, wrote Rennie, was the range of chalk hills which had no 'living waters' near its summit. The difficult stretch beyond Polhampton near Overton would require either a three mile tunnel or water to be pumped by steam engine. After passing through the hills Rennie thought there would be no difficulty in continuing the canal to Kitcomb Bridge, the point where the Andover Canal crossed the River Test and, wrote Rennie, 'I believe it may be done about the rate Mr Jessop has stated'. The Basingstoke Extension Committee requested a more detailed survey which Rennie carried out with Pinkerton's assistance in June. His report in September set out a line from the canal basin at Basingstoke to Kitcomb Bridge on the Andover Canal and between Kitcomb Mill and Old Sarum.[71]

It is interesting to note that even in 1790 there was 'some small difficulty' in getting from the basin through the town of Basingstoke where two members of the management committee lived. Mr Richard Jefferys was threatened with the loss of a corner of his pleasure grounds and Dr Thomas

Sheppard part of his garden; two small houses would also have to be pulled down; however Rennie, showing some degree of naivety, commented 'as the canal will form not only a beautiful addition to these places but a profitable one to the gentlemen, I imagine no objections will arise'.

From Basingstoke Rennie chose a gentle incline through the vale to Buckskin Farm, 2¼ miles distant, followed by a steeper ascent to the summit at Worting Coppice. The making of the canal to Kitcomb would, he thought, be 'attended with considerable difficulty' due to the poor water supply. This coupled with the crookedness' of the vale near East Dean induced him to lengthen the tunnel to 7,400 yd. Alternatively by placing the summit so high as to require no tunnel and to lock up and down each side, a steam engine would be required to pump the water.

Rennie estimated the cost with the tunnel to Kitcomb Mill at £119,077, and by locks only, at £92,080. From thence to Salisbury another £47,000 to £70,000 would be needed according to which route was chosen. The petition for the Salisbury & Basingstoke Navigation was lodged at the House of Commons in September, but Rennie's lowest estimate to Salisbury,

SALISBURY and BASINGSTOKE NAVIGATION.

NOTICE is hereby given, that a Petition will be preſented to Parliament, in the next Seſſion, for an Act to make a NAVIGABLE CANAL, which is intended to paſs through the ſeveral pariſhes and places, herein after mentioned (or ſome of them), from or near Baſingſtoke, in the county of Southampton, to or near the city of New Sarum, in the county of Wilts, viz.

The pariſhes of Baſingſtoke, Worton, Church-Oakley, Dean, Aſh, Overton, Laverſtock, Freefolk, Whitchurch, Tulton, Long-pariſh, Barton-Stacey, Chilbolton, Leckford, King's-Sombourn, Stockbridge, King's-Sombourn, Michelmerſh, Mottesfont, Lockerly, Eaſt Dean, in the ſaid county of Southampton; and Weſt Dean, Eaſt Grimſtead, Weſt Grimſtead, Alderbury, Laver-ſtock, Clarendon-Park, Milford, and Saint Martin's, in New Sa-rum, in the ſaid county of Wilts.

And that the ſaid Canal is intended to join the Andover Canal at Chilbolton aforeſaid,

SARUM, *Sept.* 3, 1790.

Many attempts were made to extend the Basingstoke Canal. One of the earliest proposals presented to Parliament was for a waterway to Salisbury and the River Avon in 1790

£139,770, excluding the necessary widening of the Andover Canal, was too much for serious consideration by the Basingstoke and nothing further was then done.

In November 1802 Phillips wrote: 'The Basingstoke Canal is improving, but until it forms a junction with some more profitable branch, it is not likely to succeed to any parties but the public.'[72] The criteria of success present in navigations which linked ports and industrial centres were missing. The Basingstoke Canal led to an area which was becoming less rather than more industrialized and so lacked the essential link with the Industrial Revolution which made for the advancement of other concerns. Ten years earlier Phillips had envisaged the canal forming part of a system to connect London with Southampton and Portsmouth. 'And if the proposed canals through the county of Hampshire should take place (which I hope I shall live to see), goods and stores of all kinds, from the magazines of London, Woolwich, etc may then be certainly conveyed to Portsmouth; thereby avoiding a long circuitous and hazardous navigation.' Indeed at that time the company was actively pursuing plans to link its canal to Newbury and the Bristol Channel and to Andover and the English Channel.

The completion of the Andover Canal in 1796 linked Andover with Southampton Water at Redbridge, a distance of 22 miles. All that remained to ensure a through passage from London to Southampton was to link Basingstoke with Andover, only 17 miles away. Such a canal, said Phillips, might be finished in twelve months and 'in time of war would save the public several millions, by accelerating expeditions and saving convoys employed between the two ports. Sometimes an expedition is deferred for want of gupowder, another time for want of guns, a third time for army necessaries. It also requires a variety of winds to proceed from the Thames to Portsmouth whereas the whole of any convoy through this canal will arrive there in three days' time.' Alternatively the junction could be made via Winchester, 18 miles distant, to which city the Itchen Navigation had been opened since 1710. Further east the building of the Arun Canal to Newbridge in 1787 had brought the navigable portions of the Wey and Arun rivers to within 15 miles of each other. A fourth possibility was Lord Egremont's proposal in July 1793 to link the Petworth branch of the Rother Navigation to the Wey at Godalming,[73] a distance of 23 miles. Any of these links would provide direct access by water for barges from London to the English Channel and so remove both the perils of shipwreck round the Foreland passage and the dangers of marauding privateers. During the next two decades war was either being actively waged with France or was not far from the horizon. It is therefore perhaps surprising that although proposals, both grandiose and minor, were many, this line of inland navigation was not completed until after the Napoleonic Wars were over; and when the link was finally forged in 1816, it was through the Wey and Arun valleys that barges were to take the slow and circuitous route from the Thames to the English Channel.

The link between Basingstoke and Winchester had first been proposed at a meeting in Southampton on 27 December 1792, with the mayor in the chair and James d'Arcy, owner of the Itchen Navigation, on the committee.[74] This idea was soon to be incorporated in the wild scheme for a canal from Salisbury to Bristol at the time of the canal mania of 1793, and the following year John Chamberlaine of Chester surveyed three lines of canal from the Kennet & Avon (whose Act had just been granted) to link Salisbury, Andover, Basingstoke and Winchester. The progress of the Salisbury & Southampton Canal revived interest in the Basingstoke link with the Itchen, and in 1796 the scheme was again put forward in the form of the London & Southampton Ports Junction Canal. The Basingstoke proprietors warily decided that it might be advantageous to forward such a junction under certain conditions and, on 4 March, Stubbs attended a meeting called by the gentlemen of Southampton. Consequently, two further surveys of the line were made; one by Joseph Hill, the engineer of the Salisbury & Southampton Canal, who estimated the cost at £127,000, and £2,540 p.a. to maintain; the other by George Smith, surveyor to the Basingstoke Canal Company, who considered that £157,566 would be required. Notwithstanding that the main argument for the link arose from the war with France, rising prices and the lack of water at the summit caused the plan to be laid aside, though in 1800 Ralph Dodd, in proposing his Grand Canal from Rotherhithe through Kingston to the River Wey (see page 69), suggested that it could be continued to the Itchen, whose locks measured 70 ft x 13 ft. As the locks on the Andover Canal were much smaller (65 ft x 8 ft 6 in.) than those on the Basingstoke (82 ft 6 in. x 14 ft 6 in.),* transshipment or the limitation of the size of through barges to those of the Andover would have proved uneconomic; alternatively, reconstruction would have been extremely costly. A link with the Itchen was therefore preferable in this respect.

In spite of the company's financial difficulties thoughts were now turned towards starting work on the six-mile branch canal to Turgis Green, whose cost was estimated at between £16,000 and £20,000.[75] In December 1797 it was stated that there was 'reason to hope that at no great distance of time, effectual arrangements may be made to begin the collateral cut, which will make a considerable addition to the revenue of the company'. In February

* These measurements are taken from *Lengths and Levels to Bradshaws Maps of Canals, Navigable Rivers and Railways from Actual Survey*, 1832.

CANAL.

AT a Meeting of the Owners of Lands and Mills, situated on the Line herein mentioned, held by Public Advertizement, at the *New Inn*, in *Overton*, on *Friday*, the 28*th* Day of *September*, 1810; for the Purpose of taking into Consideration a Project (lately circulated by MR. RALPH DODD, Engineer,) for the cutting of a NAVIGABLE CANAL, from *Basingstoke*, by *Overton* and *Whitchurch*, to join the *Andover Canal*, at *Kitcomb Bridge*.

LOVELACE BIGG WITHER, ESQ. in the Chair.

RESOLVED UNANIMOUSLY,

I.

THAT the cutting of a NAVIGABLE CANAL, in the Line before-mentioned, (if practicable,) would be Injurious and Destructive to the Rights and Interests of Private Property, beyond the Posibility of Recompence.

II.

THAT this Meeting will exert their utmost Endeavours to defeat any attempt to obtain the Sanction of the Legislature, to a Project which, without affording any solid Ground of Public Advantage, would in its Consequences prove an intolerable Nuisance to the Country, and an irreparable Grievance to Individuals. And in Order to prevent fruitless Expences and Speculations, the LAND-OWNERS feel it incumbent on them, in the first Instance, to declare and make known their Determination relative thereto.

III.

THAT a Committee, with a permanent Chairman, be forthwith appointed, with full Powers for them, or five of them, to call general Meetings of the Land-owners whenever they shall see Occasion; and to adopt such other Measures as they, from Time to Time, may deem expedient, for the Purpose of securing the Object of these Resolutions:

And that the Committee do consist of the following Gentlemen, viz.

L. B. Wither, Esq.
Wither Bramstone, Esq.
David Cunynghame, Esq.
Rev. Charles Blackstone.
John Harwood, Esq.
Rev. John Harwood.
Rev. John Smith.
William Portal, Esq.
Rev. George Lefroy.
Rev. Henry St. John.
John Portal, Esq.
John Lovett, Esq.
Thomas Streatwells, Esq.
Rev. William Harrison.
Bryan Troughton, Esq.
William Leigh, Esq.
Mr. James Crimble.
Mr. George Small.
William Bridges, Esq.
And Mr. John Twynam.

IV.

THAT the Chairman be requested to sign the Proceedings of this Meeting, and to cause the same to be published in the *Salisbury Journal*, the *Reading Mercury* and the *Hampshire Chronicle*, and also in some of the London Newspapers.

Signed
L. B. WITHER, Chairman.

Canals, like railways, were not always welcome. This notice of 1810 informs the public that a committee has been formed to resist the building of a canal to link the Basingstoke and Andover Canals

1799 Smith reported that there was a considerable fall of timber being cut in Tylney Park – nine or ten thousand loads, which if carried on the canal would provide several thousand pounds in revenue. It was thereupon agreed to make 1 mile of the collateral cut immediately from Greywell to the turnpike road on Hook Common. The cost was not expected to exceed £500. However, Bland told the proprietors in 1806 that it was the apprehension that the west mouth of the tunnel might again suffer an accident, much as happened in 1794, that the cut was proposed so that if any part of the tunnel gave way, it 'would considerably diminish the injury'. Unfortunately the committee had to report in October that they had not 'been able to bring a principal landowner to a clear and definite settlement'. Estimates were made of the differences in cost of making the cut of smaller size than the main line and consideration was given to reducing its width and using smaller barges from which goods could be removed into larger barges on entering the canal.

Cost was the main stumbling-block. To find the capital the specialty creditors were asked to remit the bonds granted them in lieu of interest on

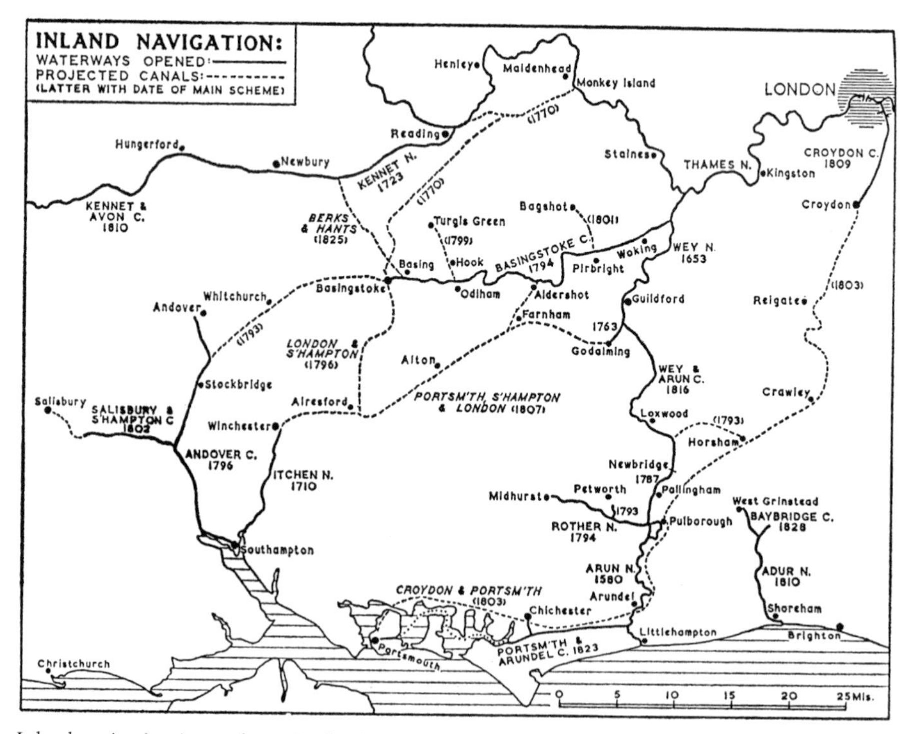

Inland navigation in southern England with dates of opening

their shares. However, although many bond holders, including both Dartmouth and Rivers, agreed to the scheme to redeem the bonds, opposition by Stubbs and one or two others was sufficient to create an impasse, since they in turn had proposed that a third Act should be obtained to raise more money. 'But,' said the chairman, 'it is to be presumed, after the full disclosure that has been made of the income of the company, which is found to be inadequate to the payment of the interest of the money already borrowed, that Parliament would not give them leave to take any further loans; or if Parliament should grant them leave, on the supposition that the trade might be hereafter so improved, as to enable the Company to pay the interest, no persons would be found to lend their money upon such precarious security.'[76] The attempt to build the branch line foundered although a shorter canal to Hook was again proposed in 1818 (see page 90).

Consideration was then given, perhaps rather surprisingly, to constructing a short branch 4½ miles long from Pirbright to the small village of Bagshot. On 24 June 1801 Dr Bland attended a meeting at the White Horse Inn, Bagshot to discuss the plan.[77] The idea behind the scheme was not only to improve the canal's water supply but to capture the four thousand or so tons of coal, corn, groceries and merchandise which went up the Thames to Staines for distribution by waggon to 'Bagshot, Sunning Hill, Chobham, Blackwater and other neighbouring estates'. Additional traffic, including chalk for manure (of which vast quantities were expected to be consumed) was estimated at not less than ten or twelve thousand tons annually, which it was suggested would afford an abundant profit to the 'adventurers'.

George Smith reported that the whole of the land was heath or common, 'in many parts of a springy texture, and evidently containing much water' and therefore well adapted to the purpose. Only one lock, close to the proposed junction with the main line, was required. Four rills which served no mill and which the chairman was informed by a Bagshot inhabitant had never failed in the driest seasons, were to provide the main water supply. The cost was estimated at £5,500. The meeting agreed that to cover contingencies £6,000 should be raised by 120 shares of £50 each. The twenty-two people attending the meeting at the White Horse Inn subscribed for fifty-one shares (£2,550), Thomas Jefferys claiming ten and Dr Bland, two. Later Prince William of Gloucester who lived at Bagshot Park, and six gentlemen in the neighbourhood subscribed a further £800. The company would have completed the subscription if funds had allowed, but being 'circumstanced' as it was, the committee proposed to take up only ten shares leaving the balance of forty-three to be taken up by individual proprietors. Although the company was authorized by its Act to make collateral cuts or branches up to 5 miles in length with the consent of the landowners, since in this instance

AN ACCOUNT OF THE TRADE AND OTHER PRODUCE, AND OF THE EXPENDITURE OF THE COMPANY,

From LADY-DAY to MICHAELMAS, 1825.

PRODUCE.	£	s.	d.	EXPENDITURE.	£	s.	d.
To Carriage of Goods from Lady-Day 1825, to Midsummer following, 4473 Tons, and to Michaelmas, 4132 Tons, ...	1719	19	3	By Salaries for six months, ...	281	4	0
To Rent of Land, Buildings, and Barges, to the same time, ...	76	6	7	By Allowance to Mr. Thomas Adams, late Wharfinger at Basingstoke, for six months, ...	30	0	0
To Dividend on £1500 Consols, six months,	22	10	0	By Labour and Materials for the same period, ...	613	18	2
To Sale and Tonnage of Chalk and Soil,	21	19	0	By Rent of Land occupied by the Canal and Works, Mr. Halsey's Annuity, and Rent of Offices and Wharfinger's House, ...	97	7	0
To Earnings of Porter, and Wharfage, ...	40	14	10	By Printing, Stationery, Postage, Insurances, Taxes, Advertisements, Expenses of General Meetings, Poor-rates, and other Incidentals, including Survey of the Line, ...	99	11	7
To Sale of Cuttings, ...	6	7	6	Balance in favour of the Company, ...	765	16	5
	£1887	17	2		£1887	17	2

Basingstoke Canal Navigation Company Report, 1825. Henry Halsey was the owner of Henly Park, Farnham, and of a mill and lands adjoining the canal at Pirbright. He was an original subscriber for three shares. In 1801 he corresponded with Dr Bland about the proposed cut from Bagshot to Pirbright which would have passed through his land (*BCN* no 45)

the cut would not belong to the company, application to Parliament was required. It was also intended to apply for additional powers to obtain water and to make extensions to Farnham and Alton. By October, two-thirds of the capital £4,000) had been subscribed, but the following month Henry Goolding completed a second survey which, while confirming the soundness of the scheme, estimated the cost at £8,125. The committee thereupon decided to abandon the project until a more favourable time arose for raising money. The scheme was never seriously revived, although it was still being advocated the following year.

Another of the difficulties which the company had to face was the problem of sailing between Richmond and the entrance to the Wey Navigation at Shepperton. A report of the Thames Navigation Commissioners in 1789 had commented on the obstructions to navigation, particularly at Laleham, Chertsey, Walton and Sunbury, which often brought fully laden barges to a halt. The commissioners commented that 'the shallows we observed in this part of the river, appear to us to be much more essential and difficult to remedy than in the upper part; and although the City have tried by breakwaters or underwater weirs, to remedy the inconvenience', there still remained many places which were greatly obstructed by shoals. As a result of the Thames Act of 1795, twenty-six pound locks were built, but only one below Boulter's, so that the busiest part of the river below Windsor suffered most at low water. A contemporary observer commented that 'the barge lies still, the bargemen are idle, or mischievous through want of employment; the owner is delayed in his freight and the freighter in his goods'.[78]

It was therefore not surprising that the Basingstoke Canal Company should have become 'peculiarly' interested in a proposal in 1802 to make a cut from Kingston to the River Wey. The company commented that the 'difficulty, danger and expenses of hauling the barges up the Thames' was proving a heavy burden on the navigation's profits and that until a safer and 'more convenient passage between London and Weybridge shall be made, it will be in vain to talk of any extension, either to Newbury, Southampton or Portsmouth, with all of which places, we shall doubtless in course of time be united'.[79]

In the late 1790s Ralph Dodd had put forward a scheme for a canal from Deptford by Clapham to Kingston with a branch to Epsom via Ewell. This project had crystallized into the Grand Surrey Canal, whose Act of 1801 authorized a line from Wilkinson's Gun Wharf at Rotherhithe to Mitcham by way of Camberwell, Kennington, Clapham and Tooting. The Grand Surrey proposed to extend the canal from Camberwell to Kingston and the River Wey, 'and thence through the Basingstoke Canal, or through the Godalming river, to Farnham, Alton, Alresford, Winchester, and by another route to Portsmouth'. The Basingstoke committee commented that the scheme seemed too extensive to be embraced in one plan, or to be executed by one set of 'adventurers', and that 'in fact that Company does not seem much in earnest in their intentions of entering upon it; at the least, the paper circulated in their name is drawn up so loosely, and gives so little information, that it is not very likely that it will lead, or rather mislead, any persons to become subscriber to it'.

The committee was right. The canal was not completed as far as Camberwell until 1810, in which year there were discussions between the Kennet & Avon and the Grand Surrey Companies about a junction between the two canals from Abingdon to Marsworth (near Tring) and from the Kennet's mouth to Isleworth.

Another scheme was for a canal from the Basingstoke Canal above lock XXIX, the top lock (except for the stop lock by Greywell Tunnel), through Farnham to near Alton, where instead of passing through the town, it would have skirted the hills between there and Alresford and gone on to join the Itchen, with a collateral branch to Portsmouth. A survey was made in the spring of 1803 by William Belworthy, 'by order of some gentlemen of Farnham and Alton', but no estimate of cost was made. Indeed there were so many schemes afoot, so much uncertainty, that the Basingstoke committee marvelled that persons could be found to engage in new adventures, knowing how fallacious such estimates almost constantly proved, and suggested that Parliament should consider obliging all newly formed canal companies to subscribe an additional sum to provide a fund for paying the management expenses, or if the scheme should fail, to pay the interest on the sum expended. Anticipating the day when debenture issues would be introduced, the report continued: 'There are few persons subscribing £100 or £1,000 who would not make their pounds guineas, this additional five per cent to be invested in the names of trustees, and the dividends allowed to accumulate for the above purposes'.[80]

In 1807 the project to link the Thames with the English Channel reappeared as the Portsmouth, Southampton & London Junction Canal. The proposed line, resurveyed by Michael Walker, favoured a barge canal 35 miles long from the Itchen at Alresford through Alton and Farnham to the Basingstoke Canal near Aldershot, or to the Wey Navigation at Godalming. The cost was estimated at £140,000; £124,400 was subscribed and 42*s* paid on each £100 share. The Basingstoke Canal Company subscribed for ten shares. A further survey by John Rennie favoured a line which joined the canal by Basingstoke Wharf rather than a little above Ash Lock which Walker had proposed and this was the line adopted when application was made for a Bill. Dr Bland commented that in view of the known utility of the undertaking he believed that no material opposition would be made to the plan. However, there were, he said,

> persons, on whom public utility has no influence, who are so miserably selfish, that to avert the smallest inconvenience to themselves, they would subject their neighbours, or their country, to the greatest peril, and as from experience we know, there never was a road, or canal Bill solicited, however useful, important, or even necessary the object, without meeting opposition, from persons of this description, the proprietors of the Basingstoke Canal, are requested to use their utmost, and earliest endeavours, in making the business known, and in procuring the attendance of as many of the members, of both Houses of Parliament, at

> the time, when the Bill shall come under their consideration, as they can, that this truely national object, may not be lost through neglect or inattention.
>
> It may have some weight with the proprietors, and excite them to greater activity, to remind them, that such a communication between London and Southampton, as will be effected by the canal now proposed to be made, formed a part of the original plan of the Basingstoke Canal. It was also a favourite object of the late Mr Pitt, and Lord Spencer, when they were in administration. They had been witness to the failure of an important expedition, attended with the loss of several transports, full of troops, owing entirely to the delay, in getting the necessary stores to the Fleet; the winds proving adverse. If this canal had been then made, that dreadful calamity would have been prevented.[81]

It was intended to overcome the problem of crossing the difficult summit by building a seven-mile railway at its highest point, but to this proposal strong criticisms were levelled on the grounds of the delay which would be occasioned by double transshipment. It was therefore decided to replace the railway by a two-mile tunnel between Ropley and Tisted and the cost was now put at £200,000.

In favour, the argument was that 'the great object will be obtained that a safe and direct communication, at all seasons of the year (without regard to wind or weather, and without danger of capture by an enemy, in the most perilous times) will be opened between the Dock Yards of Chatham and Sheerness, and the Depots at Woolwich and Deptford, and that of Portsmouth and Southampton Water (the grand rendezvous of all expeditions to the Westward).'

The scheme's opponents, however, claimed that it would cost £700,000 and even the Basingstoke Canal Committee, staunch supporters of the plan, quoted figures of £300,000 with the railway and £400,000 with the tunnel. A meeting at Alton on 8 December 1807, with Lord Stawell* in the chair, condemned the scheme on the grounds that there was insufficient water to supply the canal and that if successful it would harm the coasting trade and create unemployment among the seamen. Various other allegations were made and were partly true. It was doubted, in true altruistic fashion, whether the object of the canal was public benefit; it was, they cried, merely an attempt to get more water into the summit level of the Basingstoke Canal and to enrich George Hollis, the proprietor of the Itchen Navigation, on which only four barges were working.

* 1757–1820, third son of second Earl of Dartmouth.

A well-reasoned argument against the plan was published in 1808 entitled *Observations on the proposed Junction Canal between Winchester and the Basingstoke Canal*, in which the author pointed out that if this canal was to supply the needs of the fleet at Portsmouth, the route from Deptford would be extremely circuitous; not only would goods have to be transferred to and from the railway, but

> when we consider the small tonnage of a barge in comparison with that of such ships as are commonly used for the purpose of government, we may venture to assert that this newly recommended plan would occasion more than double the expense to the public . . . Is it not obvious to everyone who will cast his eyes on a map of the country that either a canal would be projected to run direct from Deptford (the chief deposit of Government stores) or that a cut would be made from Godalming, a distance of only 39 miles instead of 63 by the canal and Itchen navigation.

Although by the end of the year £200,000 had been subscribed, following a resurvey, John Rennie judged this sum to be less than half the sum required to link the two navigations, and in January 1809 the proposal was abandoned; a circular from the Basingstoke Canal Committee said: 'Your committee regret the necessity which led to this resolution, but the obstacles were manifold, and at this period apparently insurmountble.'[82]

These obstacles were clearly the great expense due to the hilly terrain, the probable lack of water and the problematic amount of trade, especially in times of peace. The plan was not revived; the following year saw the publication of John Rennie's ambitious proposals for the Grand Southern Canal from the Medway to Portsmouth, and when this route too had been discarded, the link was built from the Thames by way of the Wey & Arun Junction and the Portsmouth & Arundel Canals.[83]

CHAPTER SIX

THE INTENDED BERKS & HANTS CANAL (1810–27)

Opening of Kennet & Avon Canal (1810) – junction planned between Newbury and Basingstoke – conflict of interests between the Thames Commissioners and the canal companies – Berks & Hants Junction Canal Bill (1825) – many petitions both for and against – committee stage not concluded – observations by the Thames Commissioners – reply by the promotors – Bill re-presented in 1826 and lost – third application postponed – ship canal proposed (1829) – Basingstoke's advantage in remaining the terminus.

As has already been explained, the only chance of the Basingstoke Canal's economic success lay in its being able to extend its line in a direction from which trade could reasonably be anticipated. And so the opening of the Kennet & Avon Canal in 1810, linking London and Bristol via Newbury and Bath, encouraged the revival of the earlier schemes to link the Basingstoke Canal with the Kennet.[84] Ralph Dodd wrote a report to the subscribers of the Intended Junction of the Basingstoke, Andover and the Kennet & Avon Canals pointing out the decided advantages of a line from Basingstoke through Whitchurch to Kitcomb Bridge on the Andover Canal (20 miles) and from Whitchurch to the third lock on the Kennet & Avon near Newbury (14 miles). He argued that as small barges (like those used on the Andover) were the best and most used, a narrow boat canal would suffice whose cost he estimated at £187,000. He concluded that there were not many places in the kingdom in which canals can be cut that wear a brighter prospect or hold out more flattering returns of profits to the adventuring shareholder.[85]

John Rennie surveyed a line from Enborne (near Newbury) to Old Basing via Brimpton and Kingsclere Common, which envisaged a tunnel 1,500 yd long in the parish of Tadley. The estimated cost of this 21-mile link was put at £285,000. Traffic was estimated at hardly less than 30,000 tons annually to produce £6,000 p.a. and a clear profit of £4,500 p.a. after allowing for the expense of making reservoirs.[86] The Kennet & Avon Committee had 'no hesitation in unanimously recommending' that the line be adopted and a subscription list opened.[87]

Such a scheme, it was pointed out, would avoid the delays, uncertainties and insecurity of the Thames Navigation between Reading and Chertsey, but the Thames Commissioners strongly protested and published two reports[88] on the likely consequences and the arguments why their navigation should be preferred; these were claimed to be: certainty – more so than on a canal subject to frosts and droughts; cheapness – a toll of less than ½*d* per ton-mile and lower freight charges; expedition – quicker both downstream (current) and upstream (greater width) and fewer locks to pass (thirty-three compared with forty-five); and safety – only four vessels sunk since September 1798 and less risk of plunder.

Furthermore, they pointed out that their navigation, which had been considerably improved in recent years, would be imperilled through loss of traffic, that the scheme would bring Bath and Bristol only 4 miles nearer to London, and entertained 'very sanguine hopes that from a due consideration of the arguments and facts adduced, the legislature will feel its justice concerned in protecting established interests; and that a deluded public will be put on its guard against the activities of a set of speculating adventurers, who, totally unconnected with the local circumstances of the districts in which their plans are intended to operate, appear to respect no property, and regard no rights, that come in competition with their views of private gain'.

The canal proprietors, it was argued, did not have the public interests in view; it was suggested that the junction was only proposed to increase their private emoluments and attention was drawn to the Basingstoke Canal Committee's report of 18 October 1810 in which the proprietors were 'particularly called upon to exert themselves in soliciting the concurrence, and assistance of as many of the gentlemen in that part of the country through which the canal will pass as they may have access to, and when the business comes before Parliament, in procuring the attendance of as many members, as they may be able to influence'.

The scheme was also condemned by Frederick Page, owner of the Kennet Navigation which ran from Reading to Newbury, who clearly viewed with alarm the loss of his own monopoly. His strongly worded letter in the *Reading Mercury*, suggesting that the extension to Old Basing was unnecessary, prompted 'A FRIEND TO THE PUBLIC' to print a vigorous pamphlet in reply,[89] which pointed out that the public would at any rate benefit by the competition since Page would be obliged to reduce his tolls.

Another pamphlet[90] published in 1811 referred to the scheme being 'now in contemplation' and to the fact that it would open a communication for the coal, corn and iron trade into the interior of Hampshire and Surrey. The Kennet & Avon Company rather lost interest in the project, purchased the

Kennet Navigation from Page, and in 1815 considered that a link with the Grand Junction Canal would be more profitable than with the Basingstoke, since it would open up communications with the North. Its attempt to obtain an Act failed in 1819 and the depressed state of trade in 1820 caused the proposal to be dropped.

It was not, therefore, until the summer of 1824 that the attempt to form a union was revived. Francis Giles (1787–1847) surveyed a new and shorter route and, after a meeting of the Basingstoke and Kennet & Avon Committees at Speenhamland on 26 October, both companies agreed to subscribe £10,000 towards the capital of the new concern. The Somerset Coal Canal Committee also agreed to support and subscribe towards the measure and, at a special meeting held in London on 25 November, the initial subscriptions totalled over £50,000. A provisional committee was formed comprising six general subscribers, six from the two canal committees and three from landowners along the line. To pay the 5 per cent deposit, to advance money for the cost of the survey and to prepare its canal for the expected increase in traffic the Basingstoke Company sold £1,000 of its investments and suspended payments on its bond debt.

Giles proposed a canal 13 miles long from Old Basing to Midgham. The line involved 6½ miles of deep cutting and embanking, a half-mile tunnel at Tadley Hill, an inclined plane at Sherborne, 3 aqueducts, 38 bridges and some 12 or 13 locks. The summit level from Old Basing to Inwood was to be fed by the Basingstoke Canal, four small streams and the Enborne river from which a windmill or steam engine was to pump water up 50 ft into the canal through iron pipes 1 mile long. The cost was estimated at £117,600 and the objects were to improve water communications and trade between Bristol and London by avoiding the uncertain Thames Navigation above Weybridge. The anticipated traffic was listed as coal from the Somerset fields, iron, building- and paving-stone, timber, wheat and barley, oats from Wales and Ireland, as well as valuable merchandise from London and Bristol.

The Berkshire & Hampshire Junction Canal Bill was given a first reading on 17 March 1825 and was supported by no less than thirty-one petitions while opposed by sixteen. Those in favour represented the trading interests and inhabitants of the towns and cities, not only along the line of the Basingstoke and Kennet & Avon Canals between Bagshot and Bristol, but also from towns like Newport, Chepstow and Tiverton. Opposed were a number of landowners including the Earl of Falmouth and Lord Bolton, those with a financial interest in the Thames and Isis Navigations, i.e. bond-holders, traders, wharf- and mill-owners, and of course, the Thames Commissioners. The City of London stated that 'their tolls on the Thames

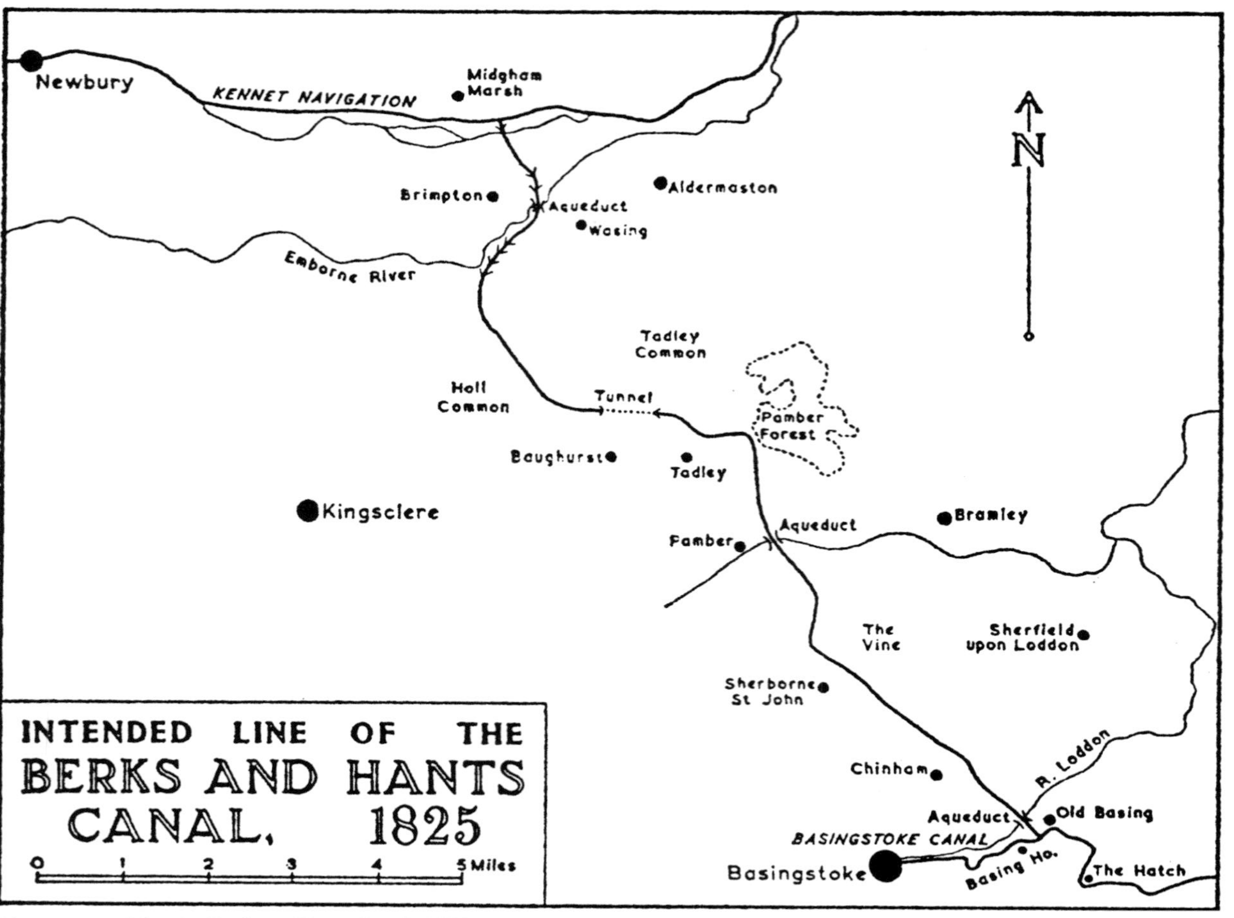

Line proposed for the Berks & Hants Canal, 1825

would be materially reduced by the trade being diverted from the river Thames on to the proposed canal'; the town of Windsor went so far as to anticipate that the river would become 'choked up from disuse' and the provost and fellows of Eton deprecated a measure which might convert their 'delightful and salubrious meadows into marshes and swamps, reeking with unwholesome exhalations'.

The Bill was referred to a committee whose chairman was Charles Dundas, a good strong supporter of the project and chairman of the Kennet & Avon as well as being the Member of Parliament for Berkshire. That this would favour the promoters was not apparent for, as was written at the time of his death, 'presiding in this committee, the recollections of his impartiality, his disinterestedness and integrity, can never be effaced from the minds of the members of it'.[91] The promoters sought to establish the need for the canal and the public benefits it would bring. However, although they produced a formidable account of barges delayed on the Thames for hours and even days, Mr Henry Joy, counsel for the Thames Commissioners, suggested that the evidence was both exaggerated and contrived, 'due to the excessive drought of the bargemen who were thus brought to, and left high and dry, at some favourite pot-house, where they dedicated a few more hours to refreshment than it was convenient to place to that score in their log-books'! So counsel insinuated it was not a want of water, but a want of beer that occasioned these delays.* The Bill's opponents also gave evidence that whereas the Thames was rarely frozen, ice had detained boats on the Basingstoke for nine, ten and thirteen weeks in recent years and that it was only in a very slight and incipient frost that the ice-boats were serviceable.

The promoters were not fortunate in their choice of witnesses. The first, Basingstoke bargeman John Gye, criticized the state of the Thames, but had unfortunately to admit that he had neither seen nor heard of the recent improvements. 'But you know for a long time past I have not been up and down the Thames; there are many present who can tell you more about it than I can.' Since he could neither read nor write, he was asked how he was able to calculate the precise time for passing the proposed junction canal. 'Ah,' he replied, 'one picks it up somehow; you get it at public houses where you get your beer.'[92]

George Long, the second witness, admitted that he would 'much rather be hanged upon the canal than die a natural death upon the river', and counsel remarked, that 'be he but half as loose in his ethics as in his evidence, there

* Mr Andras was able to instance only 30 delays out of 3,000 voyages which were recorded in his ledgers of Thames traffic.

can be no doubt but he will live to be hanged somewhere; probably on dry land'. After recounting colourful stories of the Thames floods, he was found to be oblivious of all but his more recent 'hair-breadth 'scapes i' the imminent deadly' tunnel at Greywell and the extreme difficulty of his redemption thence after an incarceration of several hours. He also told of the barge *Emma* which sank twice in the Thames, although counsel said this was 'doubtless from a kind of alacrity in sinking' for he was unable to ascertain any other reason, still less to implicate Father Thames. For, as counsel went on, the *Emma* would have equally found her way to the bottom of a canal.

What Long wanted in precision he compensated for in waggery. Questioned about barges being detained at Greywell Tunnel for two or three days for want of water, he replied: 'O, yes; a barge with only an inch of water under her is like a waggon with its four wheels locked together.'

The third witness was Bristol barge owner, Mr Shaw, 'more respectable than his two predecessors' – but he had never accompanied any of his barges except one when it broke its rudder at Shepperton due to being rammed by a Thames man astern. Although counsel for the Bill pressed him to state that it was a voyage on average of five and a half days from Bristol to London and six days to return, 'there was no coaxing him out of the awkward fact that it was made within the periods stated with great regularity'.

John Beamister told how he disliked locks since 'one cannot get a comfortable cup of tea for them'! William Heatchett voted the new works on the Thames as inefficient but couldn't place them very clearly – 'believes somebody might have put a pound at Old Windsor for a little easement', but he did give praise for the canal ice-boats and their signal service.

There was strong opposition from the landowners of whom only two had signified their consent to the project. The original line was altered to avoid Ewhurst Park but it was inevitable that some family estates would be affected; yet the Bill's opponents were hard pressed to make a good case – the canal would 'come close to Hyde End House' and the Vine, 'the hospitable home of the late Mr Chute MP for Hampshire' whose pack of hounds were still in the neighbourhood 'in spite of the threatened canal which it is apprehended will break up the hunt'. The private woods of the Earl of Falmouth and Mr Beauvoir were also likely to be infringed as well as a farmyard.

The estimates were challenged; they were undoubtedly optimistic. It was pointed out that no allowance had been made for the carrying of the spoil from the deep cuttings by the tunnel and that for 1½ miles only one boat could pass at a time. It was also doubted whether the Basingstoke Canal could handle the increased traffic since it was short of water and had a tunnel 10 in. narrower at one end than the other.

Counsel unreservedly apologized for doubting the truth of Giles' 'levels':*

> There certainly appears no ground whatever to suspect that he has taken the levels with the same looseness and inaccuracy with which he has made up the estimates. As the line is both more difficult and expensive than that suggested as more correct, it would have been impossible to impute any purposed misrepresentation, even if Mr Giles' character had not precluded the idea. Neither did we ascribe the supposed mistake to his want of competent skill as an engineer; but concluded that he must have left the levelling to other hands. But although Mr Giles is unimpeachable on this point, and in no respect to be confounded with the herd of witnesses who have stated that a river freezes more than a canal, there yet are ample reasons for distrusting that judgement upon the present question, which he has formed and delivered with so much confidence.

After referring to his £1,000 subscription to the undertaking and his likely remuneration for executing the work, Mr Joy went on:

> It is utterly vain to expect he can be an impartial witness since he is swayed by a motive far more influential than any mere pecuniary interest, in the hope of credit to be won by this achievement, the first of the kind that has yet fallen his way. He is visibly panting with all the ardour of a novice to 'flesh his maiden sword' or rather to mud his maiden spade in the canal.

And so, said counsel, warming to his theme, even if water can be procured and repairs to the Basingstoke Canal carried out, what prospect is there of any return from this monstrous proposition? He further claimed

* Giles, who was later to be appointed consultant engineer to the canal company (1829–31), had assisted John Rennie in carrying out numerous surveys including that for the proposed Weald of Kent Canal on which Rennie reported in 1802, the Portsmouth & Arundel Canal in 1815 and the Grand Imperial Ship Canal in 1825. However, as counsel for the bill's opponents asserted 'he had never been employed by that judicious engineer in the execution or superintendance of any of his works; nor has he since the death of Mr Rennie in 1812, as yet executed any canal of his own'. Indeed the attack on Giles was no more than rough justice since only the previous spring he had ridiculed George Stephenson, during the hearing of the Liverpool & Manchester Railway Bill, for his proposal to drive a railway across the marshy tract of Chat Moss. In 1830 Giles, rather than Isambard Brunel (then aged 23), became engineer of the Newcastle & Carlisle Railway and in 1834 of the London & Southampton Railway whose directors forced his resignation in 1837 for mismanagement.

> This is not a rational project for improving an eligible line of canal navigation; but is rather like many of those New World schemes, with unpronounceable names, which are so rife in these days, distinguished by such a fatuous and headlong rage for speculation that if anyone were to start a mining company in Utopia, he could presently dispose of the shares at a profit. Where is the money to come from? And what prospect is there of any return for it? Might it not just as well be thrown into the sea; and much better thrown literally into the Basingstoke Canal as in all probability there would not be water enough to cover it? The concern has long been bankrupt; its dividends are, as they have always been, naught. There is but one lingering trader upon it, and he is on the point of flitting lest his ruin should be consummated. From what quarter can a single ray of hope be expected to break in upon a scene of such utter desolation?

On 20 June, Charles Dundas reported that as there were the proprietors of land through whose property the canal would pass still to be heard, the parliamentary committee felt that all expectations of carrying the Bill through Parliament at the present advanced period of the session was 'quite hopeless'. The length of the committee stage had been 'almost unprecedented in point of duration' – more than twenty witnesses were examined and after thirty days the Bill was, so to speak, talked out by its opponents, since their evidence could not be concluded during the session. The Basingstoke Committee remained, however, optimistic, reporting that 'the measure was encouraged and supported in the strongest manner; and that the evidence brought forward by the opponents to the Bill has only added to the arguments in favour of the proposed line, both with respect to its public and private advantages'.[93] Similarly the Kennet & Avon Committee expressed the view that they 'had no doubt that had the bill not been so long detained in the Committee, it would have been carried by a great majority in the House' and cherished 'the warmest expectation of ultimate success'.[94]

The Kennet & Avon Committee recommended an increase in their subscription to £20,000 as the measure would be one of 'very great advantage' to the company, and that the 'certain increase in revenue would not be less than £5,000 p.a. and that it was not improbable that it would amount to £10,000 p.a.'

The promoters of the Bill published a reply to the Thames Commissioners' observations commenting that it was no doubt 'very satisfactory to those who do not like to hear both sides of the question' and that the management of the Thames Navigation was 'like the river itself during the summer, bound in shallows and in miseries'. At the same time they were not heedless of the criticisms levelled against the scheme and instructed both John Blackwell, the

Kennet & Avon's engineer, and Giles to carry out further surveys. Blackwell reported that the proposed line was the best which could be chosen and that the Basingstoke Canal required only certain improvements to the tunnel and the water-levels. Giles' detailed report shows, however, that the canal was leaking at Mapledurwell, the banks dry, the summit pound had sunk in places 18 in. beneath the top water-level, reservoirs needed to be constructed on the commons near Farnham, two or three bridges needed to be raised and Greywell Tunnel was 11 in. narrower at one end than the other. The Kennet & Avon's enthusiasm remained undiminished and the company further increased its subscription to £25,300 on which it advanced altogether £2,465.

The Bill was re-presented in February 1826 and was supported by twenty-four further petitions including one by the Portsmouth & Arundel Canal Company and another by the inhabitants of Cork. At the eleventh hour, meetings were held with the Thames Commissioners who agreed to withdraw their opposition on the understanding that both the canal companies would make good any loss of revenue to the commissioners if traffic between the Thames and the Kennet fell felow 25,000 tons; the Kennet & Avon agreeing to pay 2*s* per ton for every ton below 20,000 and the Basingstoke 1*s* per ton for every ton between that figure and 25,000. At the same time the Basingstoke Canal Company petitioned for a Bill to build one or more reservoirs in the region of Aldershot, Crondall, Ewshott and Crookham to improve the canal's water supply and also to raise additional capital by loan. This Bill was referred to Charles Dundas's committee on 17 February.

The promoters' hopes of success were now high but they had underrated the opposition of the landowners and the Bill was lost. Around the conference table at Newbury on 18 July sat a very glum committee whose comments about certain of the gentry were exceedingly bitter; the minutes simply record that the meeting could not 'refrain from remarking that the Bill was lost in a stage of the proceedings in which opposition is most unusual, and therefore was somewhat unexpected'. Nevertheless they still hoped by some alteration in the line 'to effect a change in the sentiments of their opponents', and members of the committee were asked to wait on the principal landowners to 'endeavour by all proper and practicable arrangements to obtain their consent to the measure'.

In September 1827 the Kennet & Avon still considered it 'desirable to renew their application for a bill' but only subject to assurances that both the Basingstoke and Wey Navigations would be put in order. Although meetings between the respective navigations took place, the Basingstoke Committee's enthusiasm appears to have waned, for it was recorded that the Kennet & Avon had re-examined the whole line 'with a view of estimating its capabilities for a junction, which that company still continue to entertain'.[95]

The Kennet & Avon, however, decided not to re-apply in the next session; negotiations for land along the line of the proposed canal were dropped[96] and the company turned its attentions to developing a railway to which it subscribed £10,000 the following year.

Not until the summer of 1828 did the Basingstoke Committee admit to their shareholders that parts of the canal had fallen into decay and 'the dangerous condition of many of the locks, bridges etc'. In May 1829 they 'deemed it unadvisable to entrust works of such magnitude to ordinary workmen', appointed Francis Giles to superintend the repairs, and, since the former working engineer had been dismissed for misconduct, pointed out that no additional expense would be incurred! Repairs were still continuing at an unabated rate two years later so it is apparent that criticisms of the navigation by the opponents to the Bill were not without justification.

The Basingstoke's annual report for 1829, circulated in the summer of 1830, commented that 'it seems almost impossible to hold out any prospect of a permanent increase of income until a junction shall have been formed between the Basingstoke Canal and one of the leading navigations in the neighbourhood: an object which the committee keep constantly in view'. And so the position was to remain.

There was a scheme in 1829 for a 'new ship canal'* for vessels of 400 tons from Deptford to Bristol by way of Sydenham, Epsom, Odiham and Devizes which would have cut through the Basingstoke in four places and the Kennet & Avon Canal in nine places.[97] At Old Stoke† near Odiham, a large basin was to be constructed from which a canal to bear ships of 700 tons was to run in nearly a south line to Portsmouth Harbour. The annual revenue from this undertaking was estimated at about £500,000 and the cost at £8 million, of which £2 million was to be offered to the public and the balance raised proportionally by each county through government loans. It was proposed to call the canal from London to Portsmouth 'George the Fourth's Canal' and that from Bristol to Stoke 'The Wellington Canal'; a third cut was to branch off to Reading, but no mention of this speculation is made in the Basingstoke or Kennet & Avon companies' reports.

Robert Mudie, writing just before the arrival of the railway, doubted whether Basingstoke suffered very much from being the terminus of the canal 'because it thereby becomes a place at which the canal-borne goods, both to and from the metropolis, are stored; and this gives a great deal of bustle and employment, and consequent profit to the inhabitants'.[98] Nevertheless, he felt

* The Grand Imperial Ship Canal project for a Suez-size navigation to link London and Portsmouth collapsed in 1827. See P.A.L. Vine, 1965, *London's Lost Route to the Sea*, chapter 9.

† Not located; possible confusion with Old Basing.

that if the canal could be carried over the summit level to the Whitchurch branch of the Test, an important line of water carriage would have been established. However, this proposal he judged impracticable, both on account of the deep cuttings required and the want of water at the summit. Indeed it is evident that if the canal had been continued southwards to join the Itchen and the link to the south coast had been executed in place of the Wey & Arun Junction Canal, it is extremely doubtful whether it would have been economically viable; and even if the Berks & Hants could have been constructed for less than £300,000, the opening of the railway from London to Bath and Bristol in 1841 would almost certainly have forced its closure by the 1860s.

Henceforth, progressive thought was turned to the projected 'steam land-carriage'. As early as 1825 the Kennet & Avon, while pressing forward for a junction canal with the Basingstoke, had considered it advisable 'in consequence of the numerous speculations entered into for making rail roads, and in particular from London to Bristol', to send its engineer into the north of England to see the operations* of several railroads and locomotive engines now in work.[99] In 1828 the Kennet & Avon turned its attention to developing the Avon & Gloucestershire Railway, which was opened for the transit of coal in 1830. In 1829 Orlando Hodgson, the agent for the Wey Navigation, was proposing to introduce steam-driven barges to enable water traffic to compete with the possible steam land-carriage, but neither the Wey nor the Basingstoke proprietors felt urged to experiment in spite of the news that plans were already afoot to build the first trunk railway in the south of England and that, with Southampton as its chosen destination, Basingstoke lay very much in its path.

* Stephenson's *Rocket* proved the possibilities of passenger rail travel at the trials held on 27 September 1825.

CHAPTER SEVEN

COMPETITION FROM LAND AND SEA (1815–32)

Basingstoke Wharf – Thomas Adams – bargemasters and carriers – the return of peace – deaths of Charles Best and Robert Bland – Richard Birnie becomes chairman – Bow Street magistrate – Samuel Attwood's diary – need for improved water supply – cut proposed to Hook (1818) – economic crisis – competition from land carriage and the coasting trade – J.R. Birnie takes a lease of Basingstoke Wharf – the Whistler case (1827) – up traffic declines – death of Sir Richard and bankruptcy of J.R. Birnie (1832).

During the early part of the nineteenth century the wharf at Basingstoke was a hive of activity. Its heterogeneous cluster of buildings extended over six acres. There were coal-pens and limekilns, saw-pits and stables, hoop-sheds and corn-stores; there were the brick and slate homes of the wharfinger and the cleaver; there were shops for the carpenter and the wheelwright, three counting-houses, and little huts for storing groceries, flour and bark. Two powerful cranes, mounds of gypsum, stacks of timber and countless wheelbarrows completed the scene. A visitor to the wharf might, in William Cobbett fashion, have commented that this appeared to be 'a grand receiving and distributing place'. Here could be observed men methodically engaged in emptying the barges newly arrived from the capital, laden with an infinite variety of merchandise, whose cargoes had to be transferred to waiting four- and six-horse waggons bound for the market towns of Wessex; beside the basin were to be found a conglomeration of carts heaped high with sacks of corn and potatoes, with fellies, underwood and staves destined for the London markets. A glance at the docks would have revealed three boat-builders at work repairing and furbishing a barge. One Constable-like figure would be leaning over with his adze to square a plank; another swinging his mallet as he caulked the timber seams with strands of oakum; while the third, long-handled brush in hand, tarred the timbers from a pot.

In the eveing the ale houses along Wote Street were filled with bargees and carters taking their ease amid the clamour of song and trumpet. All would have known Thomas Adams, the principal wharfinger at Basingstoke. His task was to ensure that the comings and goings of all and sundry to and from the wharf were well controlled; that goods were safeguarded from the risk of fire and pilferage; and that the accounts were correctly kept. Ever since the canal had been opened, he and his family had served the company in one capacity or another. After twenty-five years at the wharf, however, his infirmities had rendered him unequal to the more active duties of his situation and in 1819 he was provided with an assistant. The company pointed out that he would continue to be 'extremely useful in the counting-house and though an increase in salaries would result, the committee could not with propriety have avoided the measure'. Four years later severe illness forced him to retire, but being 'an old and faithful servant', he was generously granted an annuity of £60 a year. When Adams died at Eastrop in 1831, the committee remarked that he had 'uniformly proved himself to be an active, faithful and zealous officer of the company'.

The bulk of the trade carried on the canal from the Port of London was coal and groceries and these were discharged at every wharf between Woking and Basingstoke. Mudie drew attention to 'the fetching of coal which may come from the port of London or from some of those places on the Thames where Midland coal is brought for sale, cheaper than the sea-borne coal from London but inferior for domestic purposes'. He pointed out that the export

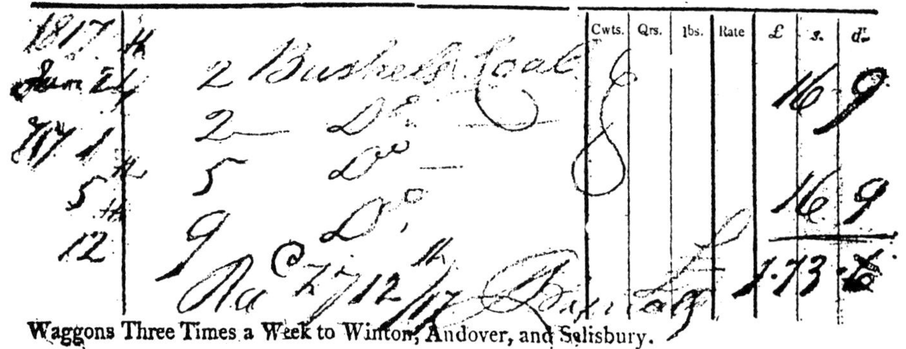

To BAGNALL and FOYLE,
BARGE-MASTERS, CARRIERS, and COAL-MERCHANTS
BASINGSTOKE WHARF.

Barges constantly working to and from Basingstoke, *and Messrs.* SILLS's, RAMSEY, *and* GRAY's Hambro' and Three Cranes Wharfs, London, *well fitted up with Tarpaulins, &c. for Conveyance of Goods which immediately on Arrival at* Basingstoke, *are forwarded by Waggons to* Sarum, Southton, Winton, Andover, Romsey, *and all Places adjacent,*—AT REDUCED PRICES

			Cwts.	Qrs.	lbs.	Rate	£	s.	d.
1817 Jun		2 Bushels Coal						16	9
		2 — Do.							
		5 — Do.							
		9 Do.						16	9
	12						1	13	6

Waggons Three Times a Week to Winton, Andover, and Salisbury.

Carrier's invoice, 1817. After the death of George Smith in 1811, the coal merchant's business at Basingstoke Wharf was taken over by Rogers & Bagnall, later to become Bagnall & Foyle of farm produce and malt were also of great benefit to Basingstoke. The predominant local traffic was timber and chalk, of which vast quantities were carried from the great pit at Odiham.[100] At the turn of the century bargemaster George Smith was the coal-merchant at the wharf. On his death in 1811 the business was taken over by Rogers & Bagnall, later to become Bagnall & Foyle, who advertised themselves as carriers as well as bargemasters and coal-merchants, operating from the wharf. Their barges – 'well fitted up with tarpaulins' – regularly ran between the London wharves of Messrs Sills, Ramsey & Gray to Basingstoke, from where goods were forwarded by waggons; three times a week to Andover, Salisbury and Winton as well as to Romsey, Southampton and all places adjacent. In 1816 Grively & Wallis, bargemasters, were to be found advertising a weekly barge service to London, by which time Wallis & Foyle and Charles Bagnall were the other rival carriers at Basingstoke.

The effects of the peace after Waterloo and the exile of Napoleon to St Helena were not so immediate as they had been in 1803. Nevertheless, at the autumn meeting in 1816 the committee drew attention to the intensive competition from not only land carriage but also 'from the facilities afforded in peace to the conveyance of goods by sea, and the fact that some considerable injury must be sustained by the canal'. Trade was also in a depressed state, although the loss of coal traffic was partially offset by the considerable fall in the price of timber.

In 1816 occurred the deaths of the company's two leading figures. Charles Best, who had been clerk for nearly forty years, died on 6 February. The committee wasted no words about his decease, mentioned only that the work for the past ten or twelve years had been mostly done in London and thought his salary might have been saved, 'the chairman and the deputy clerk having voluntarily engaged to take upon them the little additional labour'. John Birnie became clerk and the office of deputy clerk was abolished. However, four months later, on 29 June, Robert Bland died at the age of seventy-six at his house in Leicester Square, having served as chairman for over twenty years. The proprietors were informed of his decease by the report dated 5 July but any eulogy it contained must be surmised, for no extant copy has been found.

There is some evidence that, during his last years of office, Robert Bland had allowed matters to slip a little out of hand and had relied too much both on the equally aged clerk at Basingstoke and his younger deputy in London. Probably the committee was too indecisive or loathe to suggest that Bland might like to take things more easily – he obviously was – and equally obviously he did not want to relinquish either the post or his stipend of 200 guineas a year. The new chairman, Richard Birnie, whose nephew was the clerk, cut his own salary by half, thus enabling the management committee to comment that, considering the limited income of the company, the reduction in salaries 'will be found very serviceable'. Yet Bland had sought an increase as recently as the autumn of 1812 when 'there being but a small number of proprietors present' the consideration of the proposal had been deferred. Nevertheless, the company had good reason to be grateful for Bland's labours in putting its financial affairs in order. It was entirely due to his efforts that bankruptcy was averted in 1796 and that a sensible method of trading was introduced. His proposals to extinguish the bond-debt were sound. The man-midwife had shown he was not without other skills and his honorarium was only one third of that which George Stubbs had negotiated.

Richard Birnie was a Bow Street magistrate in his middle fifties when elected to office. A native of Banff, he had served an apprenticeship to a saddler before coming to work in London for Macintosh & Co, harness-makers to the royal family. Having on one occasion been accidentally called upon to attend the Prince of Wales, he gained the patronage of the prince and swift advancement in the firm to the position of foreman. Through his marriage to the daughter of a wealthy baker in Oxendon Street he secured a considerable fortune and began to distinguish himself in parochial affairs. In 1805 he was appointed a churchwarden in the parish of St Martin's and assisted in the establishment of almshouses for decayed parishioners in Camden Town. His interest in the canal company doubtless arose not only from his nephew's position as clerk, but also through his local duties bringing him into contact with Robert Bland, whose residence in Leicester Square lay only round the corner from Oxendon Street. Birnie also gave proof of his public spirit by enrolling as a private in the Royal Westminster Volunteers, in which he later became a captain. At the special request of the Duke of Northumberland he was made a JP and was at length appointed a magistrate of Union Hall. Some years later he was promoted to Bow Street. In February 1820 he not only issued the warrants for the apprehension of the Cato Street conspirators, who were plotting the assassination of the new king's ministers, but also went with the police officers to Cato Street to assist in their capture.[101] He was knighted in September 1821 on being appointed chief magistrate, to some extent as a result of the part he played at Queen Caroline's funeral the previous month, when he had read the Riot Act after the then chief magistrate had refused to do so. It was reported on 14 August 1826 that Sir Richard Birnie was too hard pressed to hear a case at Bow Street of apple stealing by two boys. *The Times* commented, 'We suspect he had no fancy for administering Mr Peel's act against a pair of urchins, one of whom appears to have been caught with no less than seven apples in his possession, which had been thrown to him by his comrade. Well, after hard pleading by the boys and their fathers for forgiveness, the two little fellows were sentenced for 14 days to the tread-mill'.

Birnie took immediate action to put the affairs of the company in order. Within weeks of taking office he and four members of the committee had gone to Basingstoke and examined the accounts. It was soon ascertained that over £4,400 was owed to the company by various traders, but that about a quarter of the debt was long-standing and the traders concerned had either died or 'become embarrassed in their circumstances'.[102] In any case he was not a man to brook procrastination. When one of the barge owners at Basingstoke got into financial difficulties, he wrote to the company's solicitors:

> I am surprised at anything bordering upon delay in the transfer of Bagnall's barges to the company – all we want is a valuation of them which by my calculation was about the amount of his debt. The company buys them at that price exchanging receipts, he for his barges, we for so much on account of his debt; surely there is nothing so simple; we did the same thing to a much larger amount with Lane in 1815 and the whole expense was the price of stamp and receipt: we are in that state which cannot support law bills. The barges next day are to be let to Bagnall.[103]

He also had the company's London office transferred to his residence at number 3 Bow Street. A further economy was made by reducing the number of shareholders' meetings. From 1817 they were held only half-yearly.

An indication of how the canal had by now become a focal point of local life is well illustrated by extracts from the diary of Samuel Attwood.[104] The Attwood family were tailors in Basingstoke. Samuel is recorded in 1838 as being a toy-dealer in London Street. His journal covers in detail a period of over fifty years from 1818 to 1870, during which time the canal is mentioned thirty-five times. The more interesting entries read:

1819	December 19	Mr Heath's waggon, loaded with stores, driven into the arm of the warfe.
1820	May 21	The 'Merry Andrew' arrived here yesterday.
1821	April 28	Went on the Basingstoke canal in Mr Cottle's boat, Easter Monday to the little Tunnel. 13 in the party. [Cottle was a printer.]
1822	July 14	Four hours to see a large party who came from Odiham in a barge with a band.
	July 28	Half day to go on a river excursion to Odiham.
	September 22	A man ran all round the common within 12½ minutes. The same person run in the warfe 50 miles in 7½ hours; his name Old Tom.
1823	August 3	Odiham water party with band here.
1824	September 5	A boy ducked in the canal supposed to be a pick-pocket.

1825	June 26	A daughter of Thomas Curtis' killed by timber stock falling on her in the warfe on Sunday afternoon while playing with other children.
	July	A great number of people died suddenly and drowned on account of the heat.
	September 10	A match of cricket played on the common between eleven men working in Basingstoke Warfe against eleven working in Odiham, Greywell and Warnborough warfs for 2/6 each man. Won in favour of the latter by nine wickets.
1826	July 2	Jonny Hillier and Joe Gillian discovered committing an unnatural crime. J. Gillian left the Town and Jonny severely ducked in the canal and at flaxpole river's almost killed him.
1827	May 20	Went to see the balloon pass near this town from Newbury and to see the new barge 'Harriett' launched.
	October 30	A lad of the name of Jordan, drowned in the tunnel this day, the employ of Mr Birnie.
1828	April 27	Two days having the rheumatism and to go to Basing in the new barge 'Salisbury' launched same day.
	August 3	Old John Hughes found drowned in the canal.
1831	May 1	Old Mr Adams the warfinger died.*
	September	Thomas Budd the smuggler died.
1835	June 7	Went with a large party to Odiham, about 45.

Attwood's record of so many deaths (see also page 128) may perhaps be linked with his family's professional connection with some of the local undertakers. One entry reads: '8 hours to attend on the petty jury and to do a little work of my own (several people dead).' Besides his references to umpiring and watching a large number of cricket matches, there are some frank admittances, '3 hours at home from the effects of being in liquor' or, 'one day to stand umpire for Basingstoke at Andover and one day to get over it'. It is also interesting to note how often the waterway appears to have been used for boating parties. This was a much commoner occurrence on the Basingstoke Canal, where there was a long lockless summit level, than on navigations such as the Wey or the Wey & Arun Junction Canal, where the time-consuming and toll-charged locks limited the number of boating excursions.[105]

Loading barges could be a tiresome process. William Smith (1790–1858), a potter of Farnborough, used the canal to move his fragile products to the capital. His grandson recalled how stowing the wares on the barges in the 1820s and '30s was a business in itself:

> a business so arduous that its details left an indelible mark on the potter's mind. What it had meant to him, his family – unborn as yet – was realised years

* Thomas Adams died at Eastrop, 6 May 1831.

> afterwards, when, on his deathbed, his wandering wits harked back and he was heard giving orders as he packed an imaginary barge. 'Come on! Let's have 'em along!', he would shout impatiently, as if at laggard labourers. During ten days of illness many hours were troubled in such a way.[106]

One of the principal problems facing any canal management was to ensure an adequate supply of water to meet the full demands of trade, and the problems it posed the Basingstoke management were little different from those which faced many other companies. The company was authorized to take water from any stream within 1,200 yd of the canal except from those running into the Rivers Loddon and Wey. The lower reaches of the canal were supplied with water from Wharfenden Lake at Frimley Green, from Mytchett Lake and Great Bottom Flash,* through both of which it passed at Ash Vale. Although the upper reaches benefited from numerous chalk springs in Greywell Tunnel, they proved insufficient during long dry summers. Consideration was therefore given to making a further attempt to begin the collateral cut from 'near the old castle at North Warnborough for about a mile to the turnpike road at Hook, notwithstanding our miserable state', in the hope of securing more water and also because 'we conceive it will secure the carriage of materials from Strathfieldsaye House, and the timber from the park to London'.† Although the scheme was approved at a general meeting in April 1818, second thoughts prevailed both in regard to the likelihood of an adequate increase in trade and supply of water. The latter was partly mitigated by the discovery of several springs close to the wharf at Basingstoke, and above Aldershot where storage capacity was increased by building a series of flashes by Eelmoor and Laffan's Plain. In spite of the total supply of water now being up to 15,000 tons weekly, shortages still continued to impede navigation.

More serious was the greatly increased competition from land carriage, as well as from the coasting trade. Towards the end of the eighteenth century the roads in Hampshire had begun to be considerably improved and in 1808 it was stated that there were no better turnpike roads in the kingdom.[107] In October 1822 the canal proprietors stated that

> the number of waggons and vans has been greatly increased upon the principal roads in the neighbourhood of the canal, and the prices of conveyance of goods have been reduced by the waggon proprietors far below the amount absolutely necessary to maintain their establishments,

* The fact that none of these is marked on any of the original plans suggests that they were natural hollows created into reservoirs when the canal was built.

† The seat of the Duke of Wellington lies 7 miles north-east of Basingstoke.

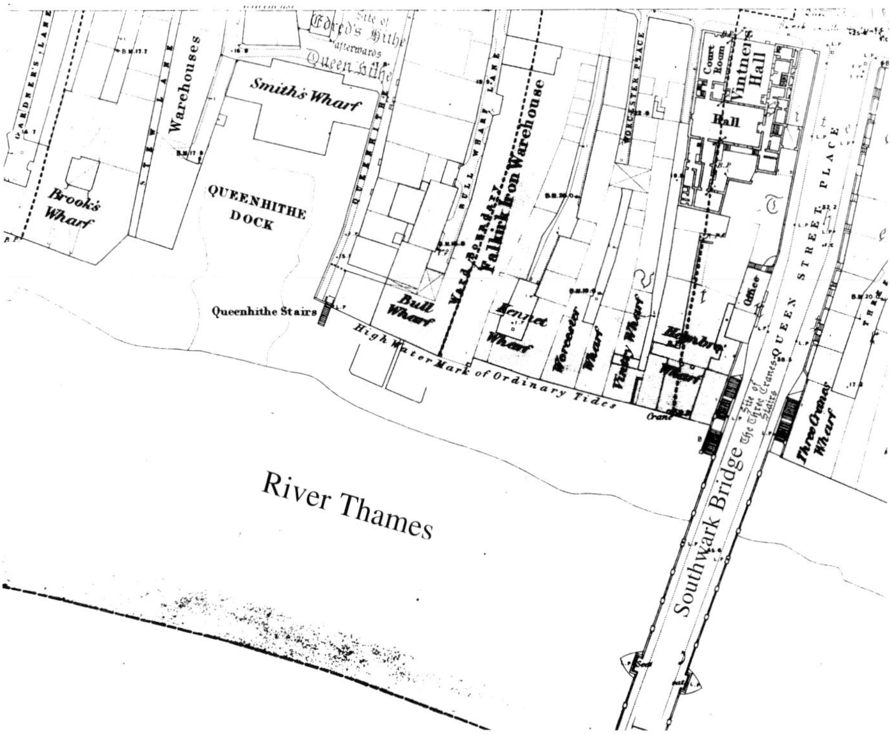

The wharves by Southwark Bridge, 1875. In the 1830s barges bound for Basingstoke were loaded at Hambro' and Three Cranes Wharf. The canal company also had an office at Hambro' Wharf until 1806. The twice-weekly fly barge to Odiham and Basingstoke, which travelled by day and night, with stops only to change horses, left from Kennet Wharf

> notwithstanding the cheapness of horse-keep. As an instance of the extreme to which the competition among waggon-masters has been carried, the Committee beg leave to state, that one hundredweight of goods is conveyed from London to Farnham by land for one shilling and sixpence, which by the canal, the same quantity must cost one shilling and threepence; and whence the difference is so slight, it may be readily imagined the land carriage will be preferred from its rapidity. In ordinary times, the expense of conveyance on land was about three times as much as by water.

Although barges took three to four days to reach Basingstoke from London, in terms of speed there was not much difference if goods were destined only from wharf to wharf; however the delay resulting from transshipment put water carriage at a distinct disadvantage when competing for cargoes to places even a short distance away from the canal.

The consequences of competition were reduced tolls, which fell from around £4,400 in 1819 to £2,394 in 1822, and a similar drop in tonnage from 22,000 to 12,000 tons. A considerable part of the trade to Alton, Andover and Alresford had been lost, and to try and win it back an abatement of 1*s* 8½*d* a ton was allowed on goods consigned to those towns. The following year saw a 25 per cent improvement in tolls and a tonnage increase of over 30 per cent, and this level was sustained until 1829 when a further improvement raised the tonnage carried to 19,000. In 1825, during the course of the hearing of the Berks & Hants Canal Bill, counsel for the Thames Commissioners, who were opposing the Bill, disparagingly stated that the Basingstoke Canal had long been bankrupt, that its dividends were, as they had always been, naught, and that one last lingering trader was on the point of flitting lest his ruin should be consummated. A grain of truth but an exaggerated assessment of the company's plight: nevertheless, during the decade, the company had shown an average trading profit of £1,300 p.a. which had never greatly exceeded £2,000 in any one year. Traffic from places other than London or on the Wey Navigation was rare, but a few cargoes did originate on the Grand Junction and the Kennet & Avon. In 1823 George Long voyaged from Bath to Odiham in a fly-boat with a load of stores. In July 1826 John Richard Birnie of Eastrop, the company's clerk, became the biggest carrier on the canal when he took over the fleet of twelve barges belonging to Wallis & Foyle. The following summer he gave up his position as clerk of the company to take up a twenty-one-year lease of Basingstoke Wharf. He was succeeded as clerk by Charles Headeach, a Basingstoke solicitor.

Chancery action against the company in 1827 caused quite a lot of trouble. Edward Whistler, a butcher of Basing, claimed ownership of land and property on Basingstoke Wharf which was contested by the company on the grounds that it was only rented. The plaintiff, however, claimed that that sum related to interest on the purchase price of £30 agreed orally by the company in 1792. Sir Richard commented on the case by saying that he was confident no such agreement existed and that he remembered in 1802 Mr Best saying to Dr Bland that Whistler wanted to buy the land and the doctor had replied: 'We do not sell – we refused to sell the banks to Sir Henry Mildmay – we may let or grant a lease, but no selling!' His view was that 'Whistler meant to purchase the land but never did so and his heirs either thought he had purchased or had such agreement, and both Dr Bland and Mr Best had a happy knack of procrastination, and like Uncle Toby's cupboard door which daily must be mended and never done'. The dispute was not finally settled until 1838 when the company obtained an indemnity from the plaintiff on payment of 50*s*.

Towards the close of the 1820s, traffic to London, which in 1824 was slightly greater in volume than that to Basingstoke, declined and by 1828 represented only half the quantity coming down. This was due to a variety of causes – the recent exhaustion of Holt Forest,* the 'great stagnation' in the timber trade, and the falling off in the carriage of flour caused by the bad harvests. On the other hand, down traffic had been improved by a considerable increase in goods destined to Salisbury, Andover, Alton and Winchester. At this time, hoops, wool and flour formed the principal up traffic while some 6,000 tons of coal were annually brought down from the Pool of London.

The passing of an Act to allow the establishment of a pitched market for corn and other goods at Basingstoke in 1829 did not produce the anticipated increase in trade, owing to a bad harvest.

The committee pointed out that the fact that the company's income was maintained, was due solely to the exertions and sacrifices made by the principal trader (the chairman's nephew, John Birnie) whose conduct demanded the warmest thanks of the proprietors. In 1830 Birnie carried 9,460 tons of merchandise between Basingstoke and the Wey Navigation, or half the total cargoes carried. His twelve barges dominated the canal. *Union* and *Commerce* were more commonly used for the coal and manure trade and *Pilot* had predominantly cargoes of beer and malt. *Alresford*, *Andover*, *Salisbury* and *Winchester* were clearly designated to make up for the canal's deficiency in terminating at Basingstoke. The patriotic *Wellington* doubtless lay peacefully at night alongside *Harriet* and *Rachel* while the *Alton Fly* was assumed to have the speed of any land carrier. Hodgson, agent of the Wey Navigation and secretary of the London & Portsmouth committee, formed to encourage traffic between these two ports, endeavoured to induce Birnie to enter the Portsmouth trade, but Birnie replied that his fleet was already fully committed.

On 29 April 1832 Sir Richard died, aged seventy-two, and John Sloper was elected chairman in his stead. Now it so happened that within only a few weeks of this event, John Richard Birnie was declared bankrupt (being in debt to the tune of £60,000, according to Attwood)[108] and had to give up the lease of Basingstoke Wharf. There was presumably some connection between the two happenings; at all events it is perhaps significant that Birnie owned a yacht which the Thames lock-keeper reported left at Weybridge back in the summer of 1831. The wharf was advertised for auction on 2 January 1833 as being 'admirably adapted and replete with the conveniences for carrying on

* Alice Holt Forest, 4 sq miles in area, near Farnham, used to supply oak timber for ship building.

(CIRCULAR.)

BASINGSTOKE, MAY 15 1832.

SIR,

WE beg to apprize you that by Deed, bearing date the 23d day of January last, **Mr. John Richard Birnie,** *of Basingstoke, Bargemaster and Coal-Merchant, assigned to* **Mr. Richard Wallis,** *of Basingstoke, sundry Debts and Sums of Money, including a Debt of £ 19 . 6 . 2 due from you to him the said John Richard Birnie, upon trust for the benefit of* **Messrs. Raggett, Seymour, and Co.** *of the Basingstoke and Odiham Bank. The Deed may be seen at our Office, in Basingstoke, and you are required* FORTHWITH TO PAY THE AMOUNT OF YOUR DEBT TO MR. WALLIS ALONE, *and not to the Assignees of Mr. Birnie, (now declared a Bankrupt,) or any other Person whomsoever.*

We are Sir,

Your obedient Servants,

Cole Lamb & Brooks

To Mr. May & Benn

R. Cottle, Printer, Basingstoke.

Notice of J.R. Birnie's bankruptcy. Consequently Birnie was forced to sell his house at Frimley and the unexpired portion of his lease of Basingstoke Wharf, described in 1833 as comprising several important warehouses, three counting houses, spacious dry sheds, saw-houses, granaries, barge building yard and stabling for about thirty horses. Birnie was a well-known figure in Basingstoke. He had worked for the company as clerk or deputy clerk for over twenty years until 1827. He was the largest carrier and in 1830 operated twelve barges carrying 9,460 tons. This represented nearly half the waterway's total traffic. His uncle was chairman from 1816 to 1832 the business of carrier, wharfinger, builder, coal and timber merchant, or any extensive business requiring space and water-communication'. At this time the property consisted of three counting-houses, dry sheds, saw-houses, granaries, a barge building-yard with steaming apparatus, wheelwright's shop, wharfinger's house and stabling for about thirty horses. The 6 acres of land were held on lease at a ground rent of £50 p.a. until 29 September 1848. Birnie's failure was a severe blow to the company – he owed it £3,754 and the accounts for 1835–6 reveal a second dividend payment in respect of this debt of only twopence in the pound. There is also an item of £501 to Birnie for the surrender of the lease and repayment of expenses or repairs. One can only surmise that the chairman's nephew had led a life far above his station, but there are no reports of Bacchanalian frolics on the barges to Basingstoke.

CHAPTER EIGHT

THE COMING OF THE RAILWAY (1830–9)

Preliminary meetings – surveys by Giles – application for a Bill – attitude of the canal company – London & Southampton Railway Act (1834) – Giles appointed engineer – Amendment Act (1837) – The Pontet case (1837) – carriage by barge of ballast and materials – line opened to Woking (1838) and Basingstoke (1839).

Horse-drawn trucks had been no threat to horse-drawn barges, but the opening of the Liverpool & Manchester Railway in 1830 established the arrival of a new form of locomotion. Charles Vancouver's report on Hampshire in 1810 had contained a heading for 'Tramroads or Iron Railways', beneath which was the sole comment, 'but of these there are not at present'. However, there was talk of a railway to Southampton as early as 1826, and in 1830 the idea was taken up in earnest.

Birnie attended a public meeting held on 6 April 1831 at the Town Hall in Southampton to consider the scheme and 'gave it as his opinion that a water communication might be made from Southampton to Basingstoke for half a million, and by thence joining the Basingstoke Canal answer nearly all the purposes of a railroad, and at much less expense. He did not consider that there was any great advantage in carrying goods with so much rapidity.' Although the scheme was regarded by many as a rather mad enterprise,[109] Birnie's opinion was not shared by the majority of the audience and it was agreed to set up a provisional committee to further the project, subscriptions being invited to defray expenses.

Giles, who was still acting as consultant engineer to the canal company, had now developed an interest in railways and had spent 1830 designing the Newcastle & Carlisle line. In 1831 he was asked by the provisional committee – mainly because he knew the lie of the land – to carry out surveys of the most suitable line between London and Southampton. However, although plans were deposited, nothing further was done until the London & Birmingham Railway had obtained its Act. The scheme did not therefore reach Parliament until 1834 when the Bill had a long hearing.

Evidence before the Lords Committee lasted from 26 May to 2 July. Among the witnesses called to prove the need for the railway were several sea captains, a Lloyd's surveyor, a tide-surveyor of Customs, and Lieutenant-

Makeshift dry dock, seen in 1966, erected above lock V at Woodham. It has now been demolished

General Sir James Gordon, the Quartermaster General, who claimed that a railway to Southampton would assist troop embarkations. Admiral Sir Thomas Hardy, one of the Sea Lords, bore testimony to the advantages which commercial vessels would enjoy, particularly in time of war and during adverse winds, of receiving and discharging their cargoes at Southampton instead of encountering the dangers of the passage from the Isle of Wight to London. Testimony was given of the great benefits which would befall fishermen, and farmers who sent 50,000 sheep annually to London markets, besides over 3,000 tons of locally killed meat. It was pointed out that driving animals caused loss of weight and value and that the health and cleanliness of London would be improved by reducing 'the nuisance of slaughter-houses'. The public was not entirely in favour. One sapient gentleman prophesied that the railway would never be made, or if made, never used; another said it would be used only for the conveyance of parsons and prawns – the one from Wincester the other from Southampton.

The railway was to run from Nine Elms through Wimbledon, Weybridge and Woodham to cross the Wey Navigation at its junction with the

Basingstoke Canal. From this point it was aligned almost parallel to the waterway for seven miles through Woking and Pirbright to Frimley, where an embankment nearly 4 miles long and 30 to 50 ft high carried the line across the heath before it cut through 90 ft of hillside and burrowed 30 ft beneath the waterway. Between Farnborough and Basingstoke the line lay well to the north of the navigation except at Basing where it came within a stone's throw of the wharf.

It is clear from the evidence available and the parliamentary papers that the company was in two minds as to what attitude it should adopt towards the railway. On 22 September 1831 the Basingstoke Canal proprietors resolved 'to adopt such measures as may be necessary for the interests of this company, in regard to the proposed railway, from London to Southampton'. The clerk, in summoning the meeting, wrote: '. . . as, in all probability, the proposed railway from London to Southampton, will either afford an opportunity for securing the best interests of this navigation, or greatly injure them, your attendance at this meeting is highly desirable.'

The standing orders of the House demanded that before a Bill could be brought in, the owners and occupiers of land through which the line was to pass had to signify their view of the scheme. It is therefore interesting to note that whereas the Wey Navigation in the person of Lord Portmore dissented, the other joint owner, the Revd Charles Langton, remained neutral; the proprietors of the Basingstoke Canal were in 1834 recorded as 'neuter at present' and John Birnie, who was now living in a house (Mylncroft) at Frimley Green, close by the canal bridge where the aqueduct was planned, assented. Certainly section LXVII of the Act would seem to confirm the view that the shareholders hoped that the canal would be bought by the railway company, and one may assume that the close relationship of Francis Giles with the directors of both the canal and railway companies increased the likelihood of such action being taken. Certainly after Giles' dismissal, the relationship between the two companies hardened considerably.

On 25 July the line was authorized to run from 'the river Thames at or near Nine Elms in the parish of Battersea in the county of Surrey, to the shore or beach at or near a place called the Marsh in the parish of Saint Mary in the town and county of Southampton'. The Act contained 217 sections. Of these section X stipulated that the width of the line should not exceed 22 yd and XVIII that the Surrey Iron Railway should be crossed by a bridge at least 15 ft 6 in. high; similarly XIX insisted that the bridge over the Wey Navigation should span the canal and tow-path in one arch and permit barges 13 ft wide, laden 7 ft 1 in. high to pass as well as horses; XX laid down that the aqueduct over the railway at Frimley should be at least 4 ft 6 in. deep and that any delay to navigation caused during construction should be compensated at the rate of 20*s* an hour;* XXI enacted that the pounds and reservoirs intersected by the railway at Pirbright should be reconstructed to supply the canal with the same quantity of water and the company compensated at the rate of £25 p.a. by way of damages; LXVII empowered the railway to purchase the canal company if both parties agreed and that if at any time after such purchase any part of the canal ceased to be used for navigation for four years and was not used by the railway, it was to be sold back to the adjoining landowners.

The principle on which the line of the London & Southampton Railway was laid down was very similar to that followed half a century earlier by the Basingstoke Canal Company – simply the avoidance of estates, especially those whose owners had shown themselves likely to offer strong opposition to the scheme. The line was therefore carried largely through a barren and desolate country, where it was said the soil was so valueless that the landowners were glad to get rid of it at any price. The author of a pamphlet written in 1834 scorned the idea that 'the traveller for pleasure, the idle tourists of the Isle of Wight or the quiet families who are coming to spend their summer at Southampton, shall all yield up the comforts of their travelling carriages and post horses and ensconce themselves behind the smoke of a steam engine'.

Francis Giles was appointed engineer from 1 September 1834 at a salary of £1,500 plus £500 expenses[110] and work was begun at Stapley Heath near Winchfield on 6 October. Progress on the line was slow. By the end of 1835 only 10 miles of line had been built and less than 30 of the 78 miles had been completed by the end of the following year. Giles had surveyed many canals in his time including the Portsmouth & Arundel and because he had planned the line of the railway to avoid as many tunnels as possible, the earthworks were on a tremendous scale – sixteen million cubic yards had to be excavated. Giles made the mistake of employing a number of small contractors working concurrently at various places on the line instead of

> those who had capital to lose and upon whom full security could be placed. It was on low and unsatisfactory estimates that the Bill was passed, and while the work was easy, while prices and pay remained depressed, while nothing extraordinary occurred, the work was done. But when any engineering novelty arose, the poor contractor was powerless. The smallest difficulty stayed him, the slightest danger paralysed him. He could not complete his contract; he lacked resources to pay the penalty; the works were often stopped; the directors as often in despair.[111]

* The Itchen Navigation was to receive compensation of only 5*s* an hour.

These failures had been foreseen by the engineers called to give evidence against the Bill before the parliamentary committees; however even during his cross-examination Giles had admitted a number of errors. Questioned by counsel about why he had not considered whether the Wey Navigation should be crossed with a brick or an iron bridge and why the clearance above water-level was shown as inadequate on the plan, Giles replied: 'I am free to confess that that particular part of the line did escape my observation; a foot or two may be excused on a line like this.'

The estimates being exceeded, the company was forced to apply for a second Act to raise a further £400,000 capital and borrow £130,000. The Manchester shareholders, who held a substantial proportion of the equity, appointed a committee of investigation and, on receiving its report, refused to subscribe additional capital unless Giles resigned. He did in January 1837.* Joseph Locke, the engineer of the Grand Junction Railway, was appointed in his stead and the work prospered.

The railway was brought into use as far as Woking on 21 May 1838. Work beyond Woking Common proceeded rapidly but not without difficulties. When the company had gone before Parliament in 1837 for its Amendment Act, seven clauses had been introduced into the Bill by the canal company which the London & Southampton did not resist for fear of being delayed in completing its line; and so it was that section XVIII allowed the railway company no more than six months to complete the aqueduct at Frimley. Failure entailed payment of damages of £100 a day until completion. This repealed the section of the earlier Act which had deterred the company from starting work, since a complete stoppage of the canal was shown to be inevitable, and the penalty clause of £1 an hour for any obstruction caused to the navigation was clearly found to be punitive. Section XIX authorized a temporary canal and tow-path to be constructed and maintained while the work on the aqueduct was in progress; section XX forbade any deviation of the line of the railway nearer the canal from the deposited plans. Section XXI compelled the railway to build a bank along the top of its embankments near the canal, sufficiently high to prevent horses on the tow-path from seeing the engines. Where the railway was 25 ft above the tow-path the bank had to be no more than 4 ft in height, but 'planted with a close furze hedge'. Section XXII required the railway to build a brick wall between 6 ft and 8 ft high (or such other fence or embankment approved by the canal company), where the

* Francis Giles often undertook more than he could manage. Sir John Rennie wrote in his autobiography that the failure of his line for the Great Northern Railway in 1844 was due to Giles's compete failure to devote to it his whole attention.

railway was built within 100 ft of the canal on or below the level of the tow-path. If built on canal property, payment of 21*s* per linear or square yard to be made for land on which the wall was built or was separated from the tow-paths. Section XXIII provided for a drain to be made at the foot of the embankments to prevent the canal being injured by water running off the embankments and section XXIV maintained a wall to protect Pirbright lock-house.

To avoid these works the directors offered their cost to the canal company, but were refused in lieu of a better offer. The railway company therefore started the walls and the Basingstoke served notice that if trains ran on the opened line before their completion, injunctions would be sought. When the following year Freeling's railway companion was published, the author commented that

> the directors were compelled to build a turf wall of considerable extent, under the pretence that their trains would frighten the Rozinantes which draw the canal barges, although they had been accustomed for months to see trains of ballast wagons running backwards and forwards many more times in the day than the trains pass; if this were opposition or intended to annoy the company or to retard the opening of the line, how dignified was the opposition! How bold the effort! What ingenuity in the mind that suggested the idea![112]

The fact that the railway was opened to Winchfield only four months after the line to Woking suggests that the rebuilding of the canal aqueduct did not constitute a major problem. The earthworks throughout the line were considerable, the Weybridge cutting totalling $2\frac{1}{2}$ miles at an average depth of 28 ft and the Frimley cutting being 2 miles long and, while averaging 42 ft, was in places 70 ft deep. Wisham, writing in 1840, stated 'there is no particular bridge or viaduct worthy of especial notice'[113] and this is substantiated by Freeling, who merely refers in passing to the 20 ft high twin arches of the 134 ft long brick aqueduct which allowed the railway to pass beneath the canal.

A Victorian idyll of peace is conjured up by the description of the waterway at Basing where Freeling's railway guide refers to the

> peaceful canal winding its sluggish way between those walls whose echoes were so often awoken by the screams of the wounded and the groans of the dying, and whose quiet was so often invaded by the spirit-stirring and busy strife of battle. Can we but regard with affection that wondrous political institution which has so long averted from our hearths the miseries of war

> and turned our energies to the cultivation of the arts of peace and the prosecution of such works as canals and railways, which must add to (by increasing the means of obtaining) the happiness of man.

In 1836 a Mr Pontet, the surviving executor of one of the bond-holders, commenced an action in the Court of Common Pleas against the canal company for the recovery of arrears of interest.[114] The committee successfully resisted these proceedings – the court deciding that having regard to the nature of the security, an action of covenant for repayment of interest did not lie against the company[115] – and hastily reminded the proprietors that the debt had at all times received its 'most anxious attention' and forthwith suggested a plan for redemption on the basis that, while interest was paid on the principal at the rate of 2 per cent p.a., bonds were to be redeemed at the rate of £50 for every £100; bonds were to be selected by lot, except that holders willing to name a lesser sum were to have priority. The committee pointedly drew attention to the fact that if the court action was persevered in, 'there will remain no alternative to prevent the party gaining a preference but that of placing the whole affairs of the company in the court of Chancery'. Nevertheless, the proposition did not receive the support of the majority of bond-holders, but at least the committee was prompted to urge that arrears of unclaimed interest up to 1824 should be claimed. The increase of trade on the canal due to the building of the railroad enabled payments of interest to be resumed, although it remained years in arrears; the 2 per cent due for 1834 not being paid until 1840. However, unclaimed interest upon the mortgage debt at the end of September 1839 was £3,365 compared with cash in hand of £6,355.

During the year ending 30 September 1836 over 33,000 tons were carried, an increase of one third in tonnage over the preceding year. The chairman drew attention in the annual report to the fact that although 'a great portion of this increase is from the materials for the Southampton Railway having been conveyed on the canal, it will be satisfactory to the proprietors to know that the general trade on the canal has much increased, and is steadily and progressively improving; and had not the competition of the road waggons much reduced the price of carriage, the revenue of the canal would have been still larger'. At this period fly-boats were running regularly. The *Maidstone Journal* on 29 September 1835 carries Wallis & White's advertisement to carry hops from London to Weyhill Fair at 2*s* 3*d* per cwt delivered. Although Weyhill was 21 miles beyond Basingstoke, the land route from London was very little shorter. An invoice dated 1839 from Wallis & White, barge owners, carriers and coal-merchants at Basingstoke Wharf, contained the information that fly-boats were loaded at Basingstoke, Odiham and Kennet Wharf, London every Tuesday and Friday.

Spanton's timber yard, Woking, *c.* 1920. Commercial carrying on the canal ceased in 1949. The last ledger entry reads, 'June 27: Gwendoline, 20 standards of timber from SS Salvas, Surrey Docks to Spantons Timber Yard, Chertsey Road Bridge Wharf, Woking. Load: 50 tons draft 3 ft 4 in.'

Traffic continued to increase as the work on the railway line progressed, reaching an all-time peak in 1838 when nearly 40,000 tons were carried, of which 33,879 were to and from the Wey Navigation. During the hectic three months before the opening of the line to Woking nearly every barge owner on the Wey Navigation had worked the canal, including those who normally never went so far afield, like Joseph Jeffs of Queenhithe, who supplied 1,000 tons of rails, George Marshall and James Webster of Godalming, James Stanton of Bramley, Charles Schofield of Kingston-on-Thames and Spencer Jupp of Littlehampton. The railway company too began running its own barge, the *Alton*, in the spring of 1837, to carry sleepers and waggons to the line.

The railway was opened to Basingstoke on 10 June 1839 and the final link with Southampton came into service on 11 May 1840. Traffic on the waterway fell immediately and from now on the Basingstoke Canal Company would be struggling not for success but for survival.

CHAPTER NINE

RIVALRY WITH THE RAILWAYS (1839–50)

The London & Southampton Railway – its threat to the canal and road carriers – toll rates reduced – less effect on tonnage than receipts – concessions granted on the Thames Navigation (1844) – opening of Guildford & Farnham Railway (1849) – resolution to sell canal carried (1850).

The early 1840s saw intense competition between the railway and the canal with both companies pursuing a policy of vigorous price cutting. Indeed, ten days before the London & Southampton Railway was opened to Basingstoke the canal company had cut its tolls, and again the following year, in order to try and retain the bulk of its heavy traffic. The Wey Navigation had done the same but the Thames Commissioners, controlling the third part of the route to London, made no move.[116] At the general meeting in November 1839, immediately after the opening of the railway to Basingstoke, the chairman John Sloper gloomily reported a fall in traffic of 12,000 tons 'attributable to the discontinuance of the conveyance on the canal of materials for constructing the South Western Railway' and not to any fall in the regular trade which the committee felt had been maintained. Nevertheless, the railway company was 'making preparations upon an extended scale to establish a formidable rivalry generally to the canal trade; but especially with regard to coals'. The chairman went on to remind the shareholders that the present juncture was one of serious difficulty and embarrassment.[117] Canal receipts had fallen by nearly a third from their inflated figure of £5,416 to £3,763 in the first year of rail competition; however, although the quantity of goods carried fell from around 34,000 to 26,000 tons, the carriage of certain items, 'particularly in hoops and timber', increased. In comparison the railway company's receipts for the second half of 1840 were £16,131 as well as £115,016 from passenger traffic.[118]

There was little optimism in the chairman's speech at the annual meeting in 1840; even the opening of the pitched market at Basingstoke, on which high hopes had been placed, had increased traffic but slightly, although on each of three consecutive days in May some 300 tons of cheese had been deposited and sold at the wharf. A minor consolation for the company's finances was the award, following an arbitration case at Guildford, of £150 for encroachments by the London & Southampton Railway on land at Woking, Pirbright and Frimley.

The railway threatened not only the canal but also the road carriers. A pamphlet of 1834, criticizing the railway scheme, said that the whole traffic between London and Southampton was at that time carried on by eight stage-coaches, four waggons per week and one barge weekly on the Basingstoke Canal.[119] Another writer, after saying that in 1835 there were in Hampshire 810 miles of turnpiked roads under thirty-six trusts, with an annual income of £30,321, went on: 'Should the opening of the railway render the conveyance of goods less costly, the number of stage waggons which will in course of time be discontinued will be much greater than in most other parts of the country where a complete line of artificial navigation is in existence. It is calculated that the number of stage waggons that will be thrown out of employment by the railway on the London & Southampton line of road will be about eighty-two.'[120] Certainly since half the merchandise carried from London to Southampton went as far as Basingstoke by water, the road carriers who consigned water-borne goods either direct to Southampton or to Winchester for transshipment on to the Itchen Navigation lost as much business as the bargemasters.

W.T. Jackman neatly summed up the reasons why the railways were so successful when he wrote: 'The railway had an air of parade and display that dazzled and tended to deceive the superficial observer. Its general aspect was that of vitality, energy and efficiency; the large trains, their promptitude of arrival and departure and the speed of the engines were all subjects of admiration, and stood out in great relief when viewed along the quiet, unseen canal and its slowly plodding barges.'[121] Railways were new and so, like all new things, they were better. So great was the passenger traffic on them that often goods could be carried at a loss with a view to withdrawing the traffic from canals.[122] It was left to only a very small minority to protest and not until nearly half a century had passed was it recognized that water as well as rail communication had a role to play in the transport system. By then, of course, it was nearly too late.

Considering the immediate tremendous superiority of rail over water transport – and in this instance the almost parallel route of the railway to the waterway precluded any chance of the navigation being able to retain traffic which could be as directly and cheaply dispatched by rail – the possibilities of maintaining the canal as an economic concern must have seemed very slight. Nevertheless, a decade was to pass before this eventually had to be seriously considered. Although the railway had most of the advantages, water traffic was still better or as well equipped and sometimes cheaper for the carriage of certain items. And so timber and coal, and materials of substantial weight but

The barge, *Red Jacket*, due to be unloaded at Spanton's Wharf in 1947. Most of the timber was loaded direct from boats discharging in the Surrey Docks. *Red Jacket* was built for A.J. Harmsworth at Berkhamstead in 1909. She was 72 ft 6 in long, 13 ft 2 in wide, and could carry 70 tons of cargo

little value, like chalk, sand and stone, continued to be carried by the waterway.

An analysis of the up traffic from the Thames to the Wey Navigation for the years immediately prior to and after the opening of the railway showed that the quantity of cargo brought by water to Basingstoke declined by 25 per cent but that the fall was very little more than 10 per cent at Odiham, which stood at some distance from the railway. For instance, in 1841 each barge owned by the traders of Basingstoke was carrying about 560 tons a year, compared with 800 in 1837, whereas Odiham barges averaged 800 tons, compared with 900.

And so the Basingstoke Canal managed to retain, and for some years increase, the amount of traffic it had carried prior to the building of the railway; whereas between 1826 and 1830 the annual freight had barely totalled 18,000 tons, between 1840 and 1844 it exceeded 21,000 tons. However, although the tonnage carried in 1848 amounted to 21,376 tons, the toll per ton received by the company had fallen from 3*s* 2*d* per ton in 1835 to 1*s* 9*d* per ton.

Workmen preparing to begin work on cleaning the channel in the Goldsworth flight of locks at Woking in 1913. Lock IX can be seen in the background with both paddles drawn. The unusual white lock gates were the result of an accidental spillage of paint on part of the gate while under construction, which necessitated painting both gates white!

In June 1843 it was reported that trade was similar to the previous year – about 23,000 tons producing a revenue of £2,312 – and that the chalk trade was reviving. As usual there was a sad note. In February 'Some evil disposed persons cut through the canal bank near Brookwood Common, thereby causing a considerable loss of water.' A reward of £25 was offered for the apprehension of the offenders, but without success.

From time to time reports were received of thefts, of deaths by drowning and of accidents. Sometimes there were lucky escapes. It was on a Monday afternoon – 8 May 1845 to be exact – when *Harriet*, one of Richard Wallis's barges, narrowly avoided being sunk while voyaging from Basingstoke to the Port of London. Bargeman Frederick Benham reported to the lock-keeper at Thames Lock that he got a shock when he was passing Ham New Mills 'not knowing or suspecting any thing of the waste gates' which had been drawn at the mills and which he couldn't see until he was upon them. 'The consequence whereof was that the aforesaid Barge came athwart the said waste gates and would of [*sic*] filled had it there not of been some person there to have shut the aforesaid gates.' Luckily Messrs Floctons men were at hand, 'as soon as we where in trouble; if not so we must have had a sunken Barge.'

By 1844 tolls had been reduced to less than half their former rates. The Wey Navigation had also halved tolls on traffic to and from the canal. In spite of these reductions, however, traders were being brought to the verge of bankruptcy. Coal merchants Wallis & Company of Basingstoke memorialized the Thames Navigation in July, drawing attention to the 'very strong and increasing opposition from the South-Western Railway Company and requesting a relaxation on the city dues of 1*s* 1*d* per ton chargeable on coal to allow us to compete with the free port of Southampton'. In November 1844 Peter Davey, the new chairman of the Basingstoke Canal Company, attended a meeting of the Corporation of London Navigation Committee to present the case for a general reduction in tolls on the Thames Navigation, for relief in respect of empty barges and for those trading to and from the Basingstoke Canal. Consequently the Court of Common Council approved the committee's recommendation that empty barges should be exempted from toll, and that laden barges should be charged only 4*d* a ton (or half toll).[123]

Not until the opening of the Guildford to Farnham Railway was the actual volume of traffic seriously reduced and the future of the company put in serious jeopardy. This line had been authorized in 1846 but not opened between Guildford and Ash Junction with a branch to Farnborough until August 1849. The remainder of the line to Farnham opened in October and was extended to Alton in 1852. The opening of Ash station (1849) took away some of the traffic from the wharf, but by granting the railway permission to reduce the size of its arch over the canal at Ash the Company received £117 compensation as well as £25 for a small strip of land.[124]

The passing of an Act in 1845 to enable canal companies to become carriers of goods upon their canals, and another similar Act two years later, prompted the shareholders to convene a special meeting to consider whether, in a last effort to rejuvenate the carrying business, the canal company should re-embark on a venture which at the turn of the century it had found uneconomic. A resolution was moved at a special general meeting held in February 1849 by one of the shareholders, the Hon. and Revd Arthur Perceval, who was also at that time chairman of the Wey & Arun Junction Canal Company. However, the proposal was defeated in favour of another: that barges should be purchased for letting to those who would otherwise be forced to cease trading.

1849 was a bad year for the company. Revenue from tolls fell by half and the tonnage carried dropped by over 8,000 tons. At the general meeting in

June 1850 Davey reported that, although every effort had been made to maintain the viability of the concern, events had proved otherwise. By economical expenditure, by reducing toll rates and by purchasing barges to let to bargemen working on the canal, costs had been reduced and attempts made to increase trade. However, the barge traders' lack of capital, the opening of the railway from Guildford to Farnham and 'the increasing reluctance to convey by water goods which are capable of a readier transit by railroad (in most cases without increased cost, and often at less expense)' combined with the withdrawal of several principal traders and carriers on the canal, had presented formidable difficulties. In addition, trade had been hampered by the long frost which had not only cost the company £30 in working the ice-boats but had held up traffic for several weeks.

The committee were indeed pessimistic. Besides stating that little prospect existed of the company's revenue being sufficiently increased to enable payment of the bond-holders' interest to be resumed – it had been suspended again in 1850 – or even of meeting the minimum expenses needed to

The aqueduct over the London–Southampton Railway at Frimley Green, 1967. It was originally built with two arches in 1838, but was extended to four when the railway track was widened in 1902

maintain the canal, they recommended the expediency of applying to Parliament for power to sell the undertaking as the only means of ensuring any return on the capital which had been so unprofitably invested. The proprietors thereupon admitted defeat and resolved that a Bill should be drafted for this purpose forthwith.

In the mid-nineteenth century some canal companies like the Andover turned themselves into railway companies, while others like the Croydon and the Thames & Medway had been taken over by them. However, no one appears to have taken much interest in the Basingstoke. Firstly, because it had proved no threat to the London & Southampton Railway; and secondly, because the route between Aldershot and Basingstoke would serve only the market town of Odiham and therefore be unlikely to generate sufficient traffic to justify a branch line. Although it was known that the South-Eastern Railway was in competition with the London & South-Western Railway for new lines in the Guildford area, it is significantly recorded in the minutes of the SER in October 1852 that they were not interested.[125]

It was at this stage that fortune smiled for perhaps the first and only time in the canal's history, and brought to the waterway a considerable quantity of very variable cargoes.

CHAPTER TEN

THE CONSTRUCTION OF ALDERSHOT CAMP (1854–9)

Establishment of military camp at Aldershot (1854) – reasons for its formation – Lord Hardinge's attention to detail – the Row Barge Inn – negotiations for sale of canal – timber and bricks by barge for the hut encampments – development of coal trade – pleasure boating.

In 1853 the government decided that the extensive heath and common land around Aldershot would make a useful area for military training. The reasons prompting this decision were not only the inadequate facilities then existing for the training of troops but also the fact that considerable compensation had had to be paid out after army exercises in more cultivated regions. In April 1854, a few days after war had been declared against Russia, the War Office confirmed the proposal to establish a permanent military camp for 20,000 men on land through which the Basingstoke Canal not only passed but would provide a cheap means of transporting military supplies.

The first reference to the possible formation of the camp at Aldershot dates from 26 September 1853, when the Commander-in-Chief, Lord Hardinge,* argued in a memorandum to the War Office that after Reigate

> the next best position for collecting troops for covering the capital and affording speedy re-inforcements for the Southern Counties, are the extensive heaths at Aldershot, Farnham and Ash. The ground is admirably suited for the assembly of a large military force, from the interior, moving to and from London, by two railways, the South Western and South Eastern . . . with an ample supply of water at all seasons. . . . I do not believe that any waste land possessing the great advantages of Aldershot, from its position towards the coast and dock yards, can be found in any other of the Maritime Counties.[126]

* Viscount Hardinge of Lahore participated in the Peninsular War but was wounded at Ligny and so was unable to fight at Waterloo. Governor-General of India in 1844, he succeeded the Duke of Wellington as Commander-in-Chief of the British Army in 1852 and was made a Field Marshal in 1855. He died in 1856 aged 71.

Flour bound for Aldershot halted below Mile Reach Brookwood because of the empty upper pound, 1915. The distance between locks XIV and XV was almost exactly 1 mile

Lord Hardinge's report and recommendations were supported by the powerful influence of the Prince Consort, who took an active interest in the army and played a decisive part in persuading the government to adopt the scheme.[127]

The Commander-in-Chief was very much concerned with the planning of the camp, too much, some people would have said. Certainly the committee which had been set up under Major-General Sir Frederick Smith to site and plan the barracks was for ever urged to make all possible speed, to seek advice from the best possible sources and, if necessary, to cite the Defence Act in order to purchase land – most of which belonged to the commoners. Agreement on some points, however, was not easily reached. Arguments ensued over commoners' rights, for which £28,000 had to be paid out, and the type of buildings to be constructed. Were they to be permanent or temporary? Thomas Cubitt, the eminent builder, favoured brick and slate, while Lieutenant-Colonel Jebb from the War Office advocated removable barracks like those he designed for the convict establishments. As the war in the Crimea fluctuated so did views on the permanence of the buildings. Finally the influx of recruits forced a compromise; brick barracks and temporary huts were both to be erected, although it was not until towards the close of the century that the huts for the militia began to be replaced.

Work on constructing the barracks commenced in September 1854 and was not completed until 1859.[128] However, in spite of the canal being so well situated to cater for the camp's requirements, the only additional barge traffic to Aldershot in 1854 was the carriage of 22 tons of fencing from London and 78 tons of earthenware pipes from Reading. Ash Green was actually the nearest railway station to the camp,* and so it was to this place that initially much of the building materials and supplies were sent for the construction of the permanent barracks and the new town which was growing up. From Ash Green a stream of horse-drawn carts was used until a light railway was built from Tongham,† on the initiative and at the expense of Mr Myers the contractor, ending just short of the Farnborough road at the rear of the main blocks of Badajos Barracks.

Amid the coming and going of troops to the Crimea, the construction of the camp progressed. Men and materials poured into the area, some to build the camp, others to develop the village into what initially was little more than a shanty town. There had originally been only three inns – including the Row Barge by Farnham Road Wharf – in the vicinity, but very soon enterprising civilians established numerous taverns and beer houses. The Row Barge, which had hitherto only catered for bargees and travellers on the turnpike road to Winchester, did, as may well be imagined, a roaring trade. It was the proud boast of the landlady that she drew a barrel of beer in fifty minutes and kept up the rate all day. The evening saw this isolated inn surrounded by crowds of soldiers and workmen, which doubtless was enough to drive the bargees away. The innkeeper, James Houghton, let out a room for the use of the Engineers' Department at 'one pound per week; coals are extra, at four shillings a week'. The inn later became the temporary headquarters of the general and his staff, and before the establishment of messes, officers were able to obtain meals there.[129] On the completion of the camp the army authorities decided that the inn, which was now on government land and which had acquired a dubious reputation through being frequented by ladies of easy virtue, should be closed down. Notice to quit was served but when the six-month period expired, Houghton showed no sign of moving and did not do so until a party of sappers began dismantling the inn and had removed the roof.

There were many delays in the construction of the camp. The initial accommodation was scheduled for completion in April 1855, but by that time

* Tongham station was opened in 1856; Aldershot station not until 1870.

† Shown on the Ordnance Survey map of 1856 but no longer marked on the edition published in 1869.

The Row Barge Inn at Aldershot stood near Farnham Road Wharf. Popular with bargees and travellers, it was demolished by the army in 1856. The turnpike road looks as if it had seen better days

only half the huts had been erected. This appears to have been due mainly to red tape and the Commander-in-Chief's close interest. In a memorandum dated 26 February 1855 to the South-Western Division Fortifications office, signed by Lord Hardinge, it is recorded that the Ordnance Department had received orders from the Secretary of State for the War Department to erect huts for 20,000 militia. Huts for 12,000 men were to be built on the south side and for 8,000 on the north side of the Basingstoke Canal (later to be known as the Marlborough and Stanhope lines). Consequently the Commander-in-Chief personally travelled down to Aldershot and met Sir Frederick and decided where the huts should be situated. The general apparently refused to delegate any major or even minor decisions to his subordinates – 'The roads round the huts will be settled by me and the Quarter Master General whenever Sir Frederick Smith will be so good as to call upon me at the Horse Guards.'

The Field Marshal added that 'the Ordnance Solicitor will have to enter into communication with the owners of the Basingstoke Canal. We shall probably require wharfs for coals, forage and various other articles.' But he

Unloading flour at Ash Lock from the narrow boats *Mapledurwell* and *Greywell* in 1916

already had, for only ten days later *The Times* became the precursor of several hundred barges to carry the paraphernalia for building and equipping the future camp. Indeed the canal company, in order to compete with the railway, had agreed to reduce its tolls on the first ten or twelve thousand tons of materials by 25 per cent.[130] The army had also been able to obtain a reduced rate of toll on the Wey Navigation of 3*d* instead of 3½*d* a ton; this rate was, however, replaced on 1 August 1855 by one of 4*d* for timber, bricks and coal, 5*d* a ton for miscellaneous items like door-frames and 6*d* a ton for hay.

Negotiations took place with a view to selling the canal to the government for the purpose of supplying the camp with water. This came to nothing but the proposal was commented on unfavourably by Stephen Leach, engineer of the Thames Navigation, who pointed out that this would result in trade on the canal being considerably reduced and that the Corporation of London had incurred great expense to provide this canal, among others, with needful facilities and in the faith that they would be maintained as navigations.[131]

The different stages of the camp's development can be traced from the changing pattern of the cargoes entered in the Wey Navigation ledgers of the

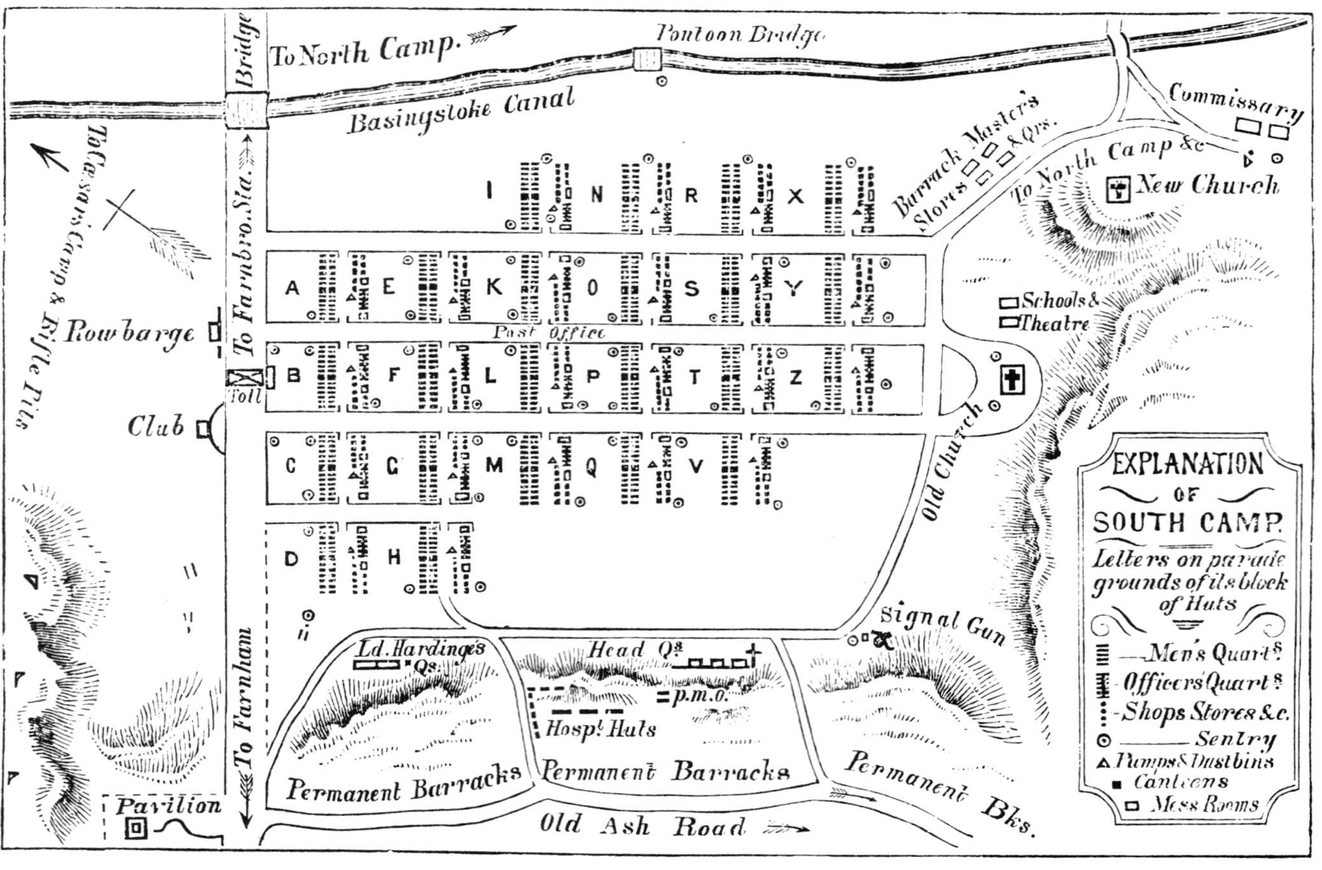

This 1856 plan of South Camp, Aldershot, shows the pontoon bridge across the canal and the Row Barge Inn, which stood by the toll-gate on the turnpike road from Bagshot to Farnham. The inn was demolished at the end of 1856

barges bound for Aldershot. Initially the principal cargo was timber and deal boarding for the huts, and bricks for the foundations but, as summer approached, slates for the roofs, paving-stones, iron pipes for guttering, and tar more frequently appeared in the bills of lading. In May 1855, 152 tons of iron bedsteads passed Thames Lock in *Defiance*. The following month *Arun* brought the first oats for the stables. In August 3¼ tons of beer relieved a local shortage and in September an 8 ton pontoon bridge was carried by *Industry* to provide communication between the two camps on the site of the present-day 'Iron Bridge' which spans Queen's Avenue. During the day soldiers were on duty to unship the bridge to let barges pass. It was closed on winter nights at 8.00 p.m. and at 9.50 p.m. in the summer; consequently every evening the open grassland between North Camp and the canal was dotted with 'red-coats' returning from the ale houses in order to catch the last 'boat'. Failure to make the bridge before closing entailed the walk up to Wharf Bridge on the Farnham road. Mrs Young, in her account of Aldershot published in 1857, relates that she met a soldier's wife on the canal bank on a Sunday who was grumbling because the bridge was not working and who had to pay an enterprising ferryman one penny in order to cross.[132]

The canal played an important part in army manoeuvres for over a century. Pontoons for bridge-making are moored by the Royal Engineers' encampment, *c.* 1910

On 13 May 1856 a less pleasant surprise for some of the troops was the arrival of the *Dispatch* laden with 1 ton of soap, but a warmer welcome no doubt greeted the *Dolphin* later that month with the delivery of the first of many pleasure boats. The arrival of 25 tons of glass in July and 23 tons of door-frames in August followed by 15 in November are curious only in that one would expect more cargoes of this nature. There is also a single, and in that respect only, mysterious entry in April 1857 for 1 ton of biscuits. Why were no more sent? Was it no more than an experimental trip, were too many eaten en route or didn't the troops like the biscuits? Such are the questions posed by ledger entries!

In July 1856 the Queen, accompanied by the Prince Consort, the Prince of Wales and the King of the Belgians, reviewed the Brigade of Guards and other troops returned from the Crimea. By this time the hutted encampments had been erected – rows and rows of wooden huts in parallel lines on both sides of the canal – and the first stage of the permanent barracks building programme was well on the way to completion. Bricks for building the permanent barrack blocks had become the predominant cargo in 1856 and

Onlookers watch the unloading of a narrow boat at the wharf below Farnborough Road Bridge, Aldershot, *c.* 1914. The outbreak of the First World War brought much additional traffic to the waterway as it was primarily used to convey ammunition and Government stores from Woolwich to Aldershot

over 7,500 tons were delivered that year to the wharf at Aldershot. This traffic continued unabated until June 1857, then ceased except for 300 tons that autumn, until October 1859 when a few more cargoes were brought up. Besides carrying building materials, the company also looked forward to conveying some of the normal requirements of the camp. In this connection over 1,000 tons of oats for the horses arrived in regular consignments from London between June and Christmas 1856 and more than 8,500 tons of coal were also delivered.

Within the space of three years 20,000 tons of building materials and commodities were brought by water to the new camp at Aldershot. Each barge had carried on average a cargo of over 40 tons, sometimes 60 tons, so that a back cargo from Aldershot was not essential for profitable performance. There was, however, a limited amount of return traffic, mainly timber and hoops, stone carted from the quarries around Alton, chalk from the great pit at Odiham, and barrels of ale from Farnham.

The Basingstoke Canal played an important role in the recreation of the Aldershot garrison. The popularity of boating led to summer regattas being held. Recognized bathing places were established and an angling club formed. Skating was eagerly anticipated when the severe frosts held the barges fast. A contemporary account of the camp in 1859[133] described rowing up the canal as a favourite pastime, 'to drink beer at a public house, where they profess to keep an "officers' room" and then to row back again'. Pleasure boats were indeed much in demand. The first arrived by barge from London in May

The canal's renown in the latter part of the nineteenth century was mainly due to the part it played in military manoeuvres, as shown in this engraving from *The Graphic*, 22 September 1888

The boathouse by Farnham Road Wharf, photographed *c.* 1910. The Dolphin had brought the first of many pleasure boats from the Thames to Aldershot in 1856 and two boathouses were soon established close to the wharf. The building beyond the boathouse was known as Boat House Café when it was auctioned in 1949. A contemporary account of Aldershot Camp in 1858 described rowing up the canal as a favourite pastime, 'to drink beer at a public house where they profess to keep an officers' room and then to row back again'

1856 and numerous entries appear in the ledgers during the succeeding summer months of these craft being brought to Aldershot. Two boathouses, Hill's and Snell's, were established close to the wharf and from these the military hired skiffs, punts and canoes. There remained (in 1968) on the west side of Wharf Bridge a derelict brick building known as Boat House Café beside which stood for many years a large dilapidated boathouse built of timber with a slate roof* – a nostalgic reminder of Victorian summer afternoons.

A new edition of *The Oarsman's Guide to the Thames* published in 1857 informed the enthusiastic boater that the trip from London to Basingstoke and back might be made in six days. It was recommended that the first night be spent at Shepperton, the second at Frimley, where the King's Head provided 'moderate accommodation' and it was counselled that 'if it be

* Demolished in 1955.

The steam launch *Una* by the carpenter's shop and forge, Frimley, *c.* 1890. This was owned by the well-known engineer Peter William Willans who was killed in an accident in 1892

intended to stop here on the way back, it may be as well to secure the necessary entertainment'. Mention was made that the pontoon bridge at Aldershot was unshipped to let craft pass. The George at Odiham, the Fox & Goose at Greywell and the Angel at Basingstoke were recommended for refreshment. To save paying 29*s* in tolls for the double journey, it was apparently in order to portage one's boat over the locks. If this was so, it was unusual, as most canal companies charged tolls whether the locks were used or not.

The construction of the camp turned Aldershot into a boom town. The site was packed with labourers in search of high wages. In their wake, and that of the soldiers, came every tart, confidence-trickster and undesirable in the kingdom. The soldiery may have been brutal; they were given little chance to be anything but licentious. Into this squalor, but separate from it, moved the court. The Queen came frequently to Aldershot to see the soldiers wheel and drill. For her comfort Prince Albert designed and had built the Royal Pavilion, which stood overlooking Laffan's Plain until the 1950s. The *Illustrated London News* described it as 'bald, cold and ugly to an extreme'. It was built entirely of wood. 'On every side there is merely a waste boggy

BASINGSTOKE CANAL NAVIGATION.

NOTICE IS HEREBY GIVEN,

THAT on and from the 12th of June, 1856, the following will be the RATES of TOLLS charged on COALS and other GOODS conveyed up and down the BASINGSTOKE CANAL from and to the RIVER WEY:—

Place	Goods	*s.*	*d.*	
Horsell and all other places below Goldsworth,	Coals,	0	6	per Ton.
	Manure,	0	4	——
	All other Goods Up and Down,	0	8	——
Goldsworth, Brookwood, and Pirbright Lock,	Coals,	0	6	——
	Manure at 1*d.* pet Ton per Mile,			——
	All other Goods Up and Down,	1	0	——
15th Lock and Frimley Wharf,	Coals,	0	6	——
	Manure and all other Goods Up,	1	0	——
	All other Goods Down,	1	6	——
Mitchett, Ash, and Farnham Road Wharf,	Coals,	0	6	——
	Manure and all other Goods Up,	1	0	——
	All other Goods Down,	2	0	——
Reading Road, and all other Places up to Basingstoke Wharf,	Coals and Stable Manure,	1	0	——
	Other Manures and all other Goods Up and Down,	2	0	——

The Tolls on the intermediate Trade (that is, those Goods which do not pass out of the Canal,) will not *exceed 2d.* per Ton per Mile, but the amount thereof will be regulated according to the particular circumstances of the Trade.

The Charge for Lockage will be discontinued, except in Special Cases.

No allowance will be made for Empty Packages, or Water in the Barges.

Every Barge is to be weighed; and the Tolls will be charged on the Number of Net Tons which may appear to be on board the Barge according to her Guage.

The Barge Owners are to give Notices of Loading and Unloading, together with an account of the Goods so Loaded or Unloaded, and must produce their Pass Tickets at every Wharf or Lock Station: and it is particularly requested that the Barge Masters will carefully examine the Guages and Tonnages entered upon the Pass Tickets, in order that any dispute as to the correctness of the same may be adjusted on the spot.

Any Barge Owner not agreeing to the above mentioned arrangements,—or not complying with the Bye Laws and Regulations of the Basingstoke Canal Company,—or fraudulently attempting to evade the payment of the Tolls,—will be charged the *Full Parliamentary Toll of 2d. per Ton per Mile on the Cargo of the Barge.*

Ample Accommodation is provided on the Company's Wharfs for the Deposit of Timber, Hoops, &c. intended to be conveyed on the Canal to London, &c. Wharfage will be charged on those Goods deposited on the Wharfs, but not finally conveyed upon the Canal: and also on Goods the Owners of which do not comply with the Rules and Regulations of the Company as to Deposit and Pitching.

By Order of the Committee of Management of the Basingstoke Canal Company,

CHARLES HEADEACH,
Clerk to the Company.

Basingstoke Canal Office, 2d June, 1856.

* *For further particulars apply as above at the Canal Company's Office, Basingstoke, or to any of the Wharfingers and Lock-Keepers on the line of the Canal.*

R. COTTLE, PRINTER BASINGSTOKE.

Basingstoke Canal toll rates, 1856.

In 1856 the opening of Tongham station, 3 miles from Aldershot, forced the canal company once again to reduce the toll on coal. The charge of 6*d* a ton to Aldershot was less than one fifth of the maximum toll authorized by the Act over eight years before

BASINGSTOKE
CANAL
NAVIGATION.

TAKE NOTICE,

That all Persons Trespassing on these PLANTATIONS under any pretence whatever, or committing *Damage, Injury,* or *Spoil thereon, will be* forthwith prosecuted.

By Order of the Company of Proprietors of the Basingstoke Canal Navigation.

CHARLES HEADEACH,
Clerk to the Company.

Basingstoke, 1st October, 1858.

R. Cottle, Printer, Basingstoke.

Notice to Trespassers, 1858.

The warning notice reflects the efforts made by the canal company to increase its income by planting firs and other trees on land bordering the canal, 1858 moor, dreary and repellant in its aspect. . . . In the foreground is a long piece of muddy water called the Basingstoke Canal.'[188]

The coming of the camp certainly changed the face of Aldershot. Before, it was one of the most pleasing and picturesque hamlets in Hampshire, with a population in 1851 of only 875. Ten years later it had become a great military station with a population of 7,755 and with 13,000 troops in the vicinity. The canal company's records for this decade are missing and while the canal's trading activities can be traced from the account books of the Wey Navigation, the details of the negotiations between the company and the military authorities can only be deduced from the pattern of trade. Clearly rail was preferred to water transport wherever practicable; nevertheless the need for pushing forward the building of the camp with all possible speed resulted in a fair proportion of trade passing to the waterway. In the short term the canal company had benefited from the tolls for carrying materials to build the camp, from receiving a certain amount of compensation for damages and from the purchase of adjoining land for wharves and other purposes. In the longer term it held two advantages over rail transport; firstly in moving military supplies direct by water from the garrison at Woolwich, and secondly in carrying munitions. The London & South-Western Railway remained equally keen to obtain the coal traffic and a price war was inevitable, with the army changing carriers as one or other offered more favourable rates of carriage.

Nevertheless, the amount of traffic passing annually to and from the Wey Navigation on to the canal during the second half of the decade was double that of the preceding half and reached 27,500 tons in 1856 compared with 8,500 tons in 1852. In addition, substantial local traffic was created around Aldershot, which caused the navigation to be used as much as at any time in its history. Although the quantity carried almost equalled the best of the pre-railway years, competition with the railway prevented the canal company from earning little more than was enough to maintain the waterway. The building of Aldershot Camp had saved the canal from a premature demise, but even so the future of the navigation remained insecure. This was soon to be proved the case.

CHAPTER ELEVEN

RISE AND FALL (1860–6)

Increase in traffic but receipts dwindle – a bargemaster's complaint – the sinking of the Horsham (1861) – action against Brassey & Ogilvie fails – loss of Aldershot coal and manure trade (1865) – committee appointed to consider the navigation's future – deaths by drowning – in Chancery (1866) – appraisal of the Basingstoke Canal – share and bond prices.

William White, writing about the canal in 1859, stated that the shareholders had never received a dividend and that the arrears of interest on the bonded debt of £32,000 now amounted to £105,000.* However, he added, 'during the last few years it has been under the management of men of business habits who have been enabled to pay some little interest to the bond-holders.'[134] The establishment of the military camp at Aldershot had certainly given an impetus to trade, and during the five years 1860–4 traffic on the waterway was as busy as at any period of its history, averaging annually around 30,000 tons. Unfortunately for the canal company, however, the tolls received amounted to no more than 10½*d* per ton, which represented little more than ½*d* per ton-mile, or barely a quarter of the maximum rate laid down by the Act eighty years before.

Nevertheless, by 1861 the trading position had become more stabilized, even if the general outlook was far from promising. Although the completion of the new wharf at Ash Lock ensured an increased demand for coal for the government gasworks[135] and the trade from London had improved slightly, due to the arrangements concluded with the government for all military stores from the Tower of London and Woolwich Arsenal to be carried to Aldershot in government barges instead of by railway, it was clear that additional revenue must be sought. Accordingly, toll rates on upwards goods were increased in 1862, the company feeling that what little traffic was left was unlikely to show any marked deterioration. If it did, then the only solution was to wind up the company.

This policy did not meet with the approval of some of the bargemasters, 'so them that suggested that was not considering the wellfair of the canal nor

* Not verified.

Pirbright lock-house and lock XV, *c.* 1915. In 1890 the London & South-Western Railway built a steel girder viaduct over the canal above Pirbright Wharf to carry a branch line from Brookwood to serve the newly transferred National Rifle Ranges at Bisley. In 1916–17 the army constructed extensions to serve the camps at Deepcut and Blackdown. This latter service ceased in 1921 and the line to Bisley was closed in 1952. The viaduct was demolished by British Rail in 1980

them that works on it'. Thus wrote James Wilkins of Ash in a letter to the Wey Navigation in April 1864,[136] which paints a sorry story of the declining trade. Apparently he was being charged 6*d* a ton on the Wey and 1*s* a ton on the canal for freights of Thames sand bound for Aldershot.

> I rote to the canal company and told them I could not do it unless they lowered there tolls to 5*d* per ton on sand to Aldershot; the river Wey ought to lower their to 2*d* a ton also. I told them I could get 2 or 3 freights more to Aldershot if they would lower thare tolls. But I cannot do it unless the tolls is lowered for I cannot compeat with the Railway and I beg to inform you that there is hundreds of tons of goods every year goes by railway to Aldershot then we should get a great share of if the tolls was lowered. I think the companys ought to be more leanient to us when they know that the Railways are working cheap and the trade against the canal. I dont like

> to see it and we lying still half of our time is no good to the company and it hurts us very much for as soon as coals is dear we have nothing to do. I hope you will bare in mind that Aldershot is a place surrounded with railways and all likeleward of bean another before long which will crush us altogether if thare is not something don for us and you know that our canal is a bad one for warter and the expense for crossing to Aldershot is grate and we cannot get aney more per tons to Aldershot than we do to Goldsworth.

An incident on the Thames, which had unfortunate repercussions, occurred in the spring of 1861. The company's barge *Horsham* loaded with coal was caught foul in the chain of another barge driving piles for the new railway bridge at Battersea. The *Horsham* swung about and struck her forward bow against the piles and sank. No one was injured, part of the cargo was saved and the barge refloated. The company alleged negligence on the part of the contractors and claimed £200 damages, mainly in respect of consequential loss. The case was set down in the Queen's Bench Division as *Basingstoke Canal Navigation* v. *Brassey & Another** and referred by mutual agreement to an arbitrator. The hearing took place at the Westminster Palace Hotel on 20 February and took the customary form of a running-down action with the defendants claiming negligence on the part of the plaintiffs. On 4 April 1862 Mr Hamilton Fulton held that the company was not entitled to recover and awarded costs against it. After the company had had second thoughts about having this decision set aside, it found that the legal costs had totalled nearly double the amount at issue. Shades of *Jarndyce* v. *Jarndyce*!

The establishment of the camp at Aldershot had brought mixed blessings. Soldiers are no respecters of property and army exercises had led to damage to canal property both wanton and accidental. Reference was made in the chairman's speech at the annual meeting in 1861 to the various claims laid before the War Office which 'had been fully discussed and the point in dispute referred', with the view of arriving at an amicable adjustment.[137]

In 1865 the canal's financial position worsened critically when coal supplies for Aldershot Camp reverted to the railroad and the camp horse-manure trade ceased. The great shortage of water during the summer had also operated very unfavourably against the general trade upon the navigation.

On 18 December 1865 a special meeting was held at Grays Inn Coffee

* The firm of Thomas Brassey and Alexander Ogilvie was building the bridge under the West London Extension Railway Act, 1859.

House, Holborn (meetings had been held either here or at the Law Institution in Chancery Lane since the 1840s) to discuss the company's predicament. It was a rather sad occasion. After informing the bond-holders that the receipts were not equal to the expenditure, the chairman Peter Davey said there was little prospect of increasing the traffic. Their land contiguous to Aldershot Camp was left in a sandy state, 'owing to the evolutions of the troops'; as a result of wind and rain the canal had become blocked on several occasions and had had to be cleared at their own expense. They had received some assistance from the government but latterly they had been refused further cooperation so that they were now left powerless to work the canal or to remove obstructions for want of funds.

The meeting went on to consider the use of the canal as a water supply for Aldershot Camp – the water from the canal being apparently thought superior to the camp supply – and also to examine the possibilities of turning it into a railway. On this latter point Joseph Woods, the railway engineer, had been consulted and had ventured the view that the financial return would bear no relative proportion to the cost. At length the resolution was moved proposing the adoption of measures for closing the canal, or its conversion into property of a remunerative character.[138] This was unanimously carried. Alternative suggestions to continue it as a reservoir for water were unsuccessful.

Although the canal's trading activities were becoming less and less regular, the navigation meanwhile maintained a certain notoriety in respect of the number of people drowned in its waters. The diary of Samuel Attwood recorded no less than thirteen cases of drowning in the canal around Basingstoke between 1825 and 1868. His rather blunt, factual entries are made without comment, though a possible cause of these tragic events can in some instances be deduced. In July 1825 'a great many were drowned due to the heat'. On 30 October 1827 a boy called Jordan, employed by Birnie, was drowned in the tunnel. A shopman, Charles Reeves (1834), Thomas Wainwright's grandson (1846) and John Spanshott (1853), late of Salisbury, found watery graves. In February 1854 the not-so-youthful Attwood went sliding six times but didn't fall in. Both Mrs Gillian of Basing and fifty-year-old John Railkin perished in the canal in 1858, the latter with a watch and £1 2*s* ½*d* in his pockets. The following year Henry Dixon was drowned when returning from Odiham Races and in 1862 William Hillier met the same fate at Greywell Tunnel after a fire at his premises. George Hall, a wheeler, and George Clarke, 'the drunken tailor', lost their lives in 1867, the former being found 'in the new berth near the pound'. Attwood's last entry relating to the canal refers to a man by the name of Samuel John Pask, lately a landlord at Battersea, who was found drowned on 7 August 1868. Doubtless there were many other tragedies. The *Surrey & Hants News* averred that no less than ninety men wearing Her Majesty's uniform had been discovered drowned since the camp had been founded ten years before, and demanded:

> At whose door shall this ghastly company of soldiers stand? By whose neglect and improvidence were these ninety destroyed? Shall the martyrdom of this company be charged against the Government or against the trustees of the Basingstoke Canal? We most earnestly commend this cause to the serious attention of the 'Society for Propagating the Gospel among the Patagonians', believing honestly that they might evidence their faith in the mission to that far-off sphere by the 'works' necessary to allow, keep and maintain a mile or two of posts and chains on the banks of this canal. A little sympathy towards the soldiers who serve their Queen and country in far-off climates would not be out of place for those individuals whose fine philanthropies carry them so far. But, perhaps, distance lends enchantment to the view. Perhaps too, these worthies, like other children, prefer coloured subjects; for them the 'pure uncoloured' white – bodily, socially, and spiritually – possesses no feature of interest.[139]

The accounts for the year ending on 25 March 1866 showed a fall of over 13,000 tons in the cargo carried and a drop of 25 per cent in toll receipts compared with the previous year. At the general meeting held on 4 June 1866 it was resolved that application should be made to the Court of Chancery to liquidate the Basingstoke Canal Navigation Company. A winding-up order was made on 23 June and Frederick Whinney was appointed official liquidator. Those, however, who thought the closure of the canal imminent were to be proved wrong, as events will show.

At this point of time, with the original company of proprietors going into liquidation, it is perhaps appropriate to look back over the past seventy years and assess what the building of the Basingstoke Canal had achieved. Financially it had, of course, proved a disaster to both shareholders and bond-holders. The value of the original £100 shares had soon fallen. Prices realized at the sale of thirty £100 pound shares in London on 18 February 1795 ranged from £51 to £77, the average being just under £72. By 1800 'the biddings barely amounted to £30', in 1818 to less than £10[140] and in 1825, during the committee hearing of the Berks & Hants Canal Bill, it was said that their price had risen from a nominal £5 to an actual £12 or £15; in 1830 two shares changed hands at £8¼ and, although in 1834 reference was made to their value as being 'not above £5',[141] a price of £7½ was recorded in 1849. Likewise, in 1795 the £100 bonds with arrears of interest sold for £90, but five years later fetched less than £60. Although payments of interest at a rate of 2 per cent were resumed in 1808, they were always several years in arrears.

PARTICULARS

OF

Thirty Shares in the Bafingftoke Canal Navigation;

OF

One Bond for the Sum of Six Hundred Pounds,

AND OF

Three other Bonds for the feveral Sums of One Hundred Pounds each,

WITH INTEREST,

Secured by the Company of Proprietors of the faid Navigation:

Which will be Sold by Auction,

BY MR. WILLOCK,

At the Rainbow Coffee-Houfe, in Cornhill, London,

On WEDNESDAY the 18th of FEBRUARY, 1795, at TWELVE o'Clock,

IN THIRTY-FOUR LOTS.

LOT I. ONE HUNDRED POUND SHARE in the NAVIGABLE CANAL from the Town of *Bafingftoke*, in the County of *Southampton*, to communicate with the *River Wey*, in the Parifh of *Chertfey*, in the County of *Surrey*, and to the South Eaft Side of the *Turnpike Road*, in the Parifh of *Turgifs*, in the faid County of *Southampton*.

☞ *This Navigation has been recently compleated and opened for proper Barges, and it is expected that the Tonnage, which is already very confiderable, will rapidly increafe, and produce great Advantage to the Proprietors.*

LOT II. ANOTHER SHARE in the faid CANAL.
LOT III. ANOTHER SHARE in the faid CANAL.
LOT IV. ANOTHER SHARE in the faid CANAL.
LOT V. ANOTHER SHARE in the faid CANAL.
LOT VI. ANOTHER SHARE in the faid CANAL.
LOT VII. ANOTHER SHARE in the faid CANAL.
LOT VIII. ANOTHER SHARE in the faid CANAL.
LOT IX. ANOTHER SHARE in the faid CANAL.
LOT X. ANOTHER SHARE in the faid CANAL.
LOT XI. ANOTHER SHARE in the faid CANAL.
LOT XII. ANOTHER SHARE in the faid CANAL.
LOT XIII. ANOTHER SHARE in the faid CANAL.
LOT XIV. ANOTHER SHARE in the faid CANAL.
LOT XV. ANOTHER SHARE in the faid CANAL.
LOT XVI. ANOTHER SHARE in the faid CANAL.
LOT XVII. ANOTHER SHARE in the faid CANAL.
LOT XVIII. ANOTHER SHARE in the faid CANAL.
LOT XIX. ANOTHER SHARE in the faid CANAL.
LOT XX. ANOTHER SHARE in the faid CANAL.
LOT XXI. ANOTHER SHARE in the faid CANAL.
LOT XXII. ANOTHER SHARE in the faid CANAL.
LOT XXIII. ANOTHER SHARE in the faid CANAL.
LOT XXIV. ANOTHER SHARE in the faid CANAL.
LOT XXV. ANOTHER SHARE in the faid CANAL.
LOT XXVI. ANOTHER SHARE in the faid CANAL.
LOT XXVII. ANOTHER SHARE in the faid CANAL.
LOT XXVIII. ANOTHER SHARE in the faid CANAL.
LOT XXIX. ANOTHER SHARE in the faid CANAL.
LOT XXX. ANOTHER SHARE in the faid CANAL.

LOT XXXI. ONE BOND for the principal Sum of SIX HUNDRED POUNDS, of the *Company of Proprietors* of the *faid Navigation*, with *Intereft* for the fame, at *Five per Cent. per Annum*, punctually paid, Half Yearly, at the Proprietors' Office in *Suffolk-Street, Charing-Crofs, London.*

LOT XXXII. ANOTHER BOND for the principal Sum of ONE HUNDRED POUNDS, of the faid Company, with *Intereft* for the fame, at *Five per Cent. per Annum.*

LOT XXXIII. ANOTHER BOND for the principal Sum of ONE HUNDRED POUNDS, of the faid Company, with *Intereft* for the fame, at *Five per Cent. per Annum.*

LOT XXXIV. ANOTHER BOND for the principal Sum of ONE HUNDRED POUNDS, of the faid Company, with *Intereft* for the fame, at *Five per Cent. per Annum.*

Printed Particulars, with Conditions of Sale, may be had at the Crown Inn, at Bafingftoke; at the Rainbow Coffee-houfe, Cornhill; Baptift Coffee-houfe, Chancery-Lane; and of Mr. WILLOCK, No. 25, Golden-Square, London.

Particulars of Basingstoke Canal shares sold by auction in 1795. Although the canal was now operating, the average price realized was less than £72 for each £100 share

In 1805 the company had begun to repurchase bonds.

> As similar opportunities may occur, the proprietors will consider whether it may not be proper to empower the Committee to purchase bonds, offering upon terms so highly advantageous. Besides some circumstances, making this eligible, which cannot with propriety be mentioned in a public paper of this kind, it is obvious, that by such purchases, the bonds that remain will be improved in their value; it will also facilitate the carrying into execution, the plan for reducing bonds, long since laid before the proprietors.[142]

Their price rose slowly from £40 in 1805 to £46½ in 1810. The rebuilding of Odiham Wharf delayed their repurchase, but between 1817 and 1821 the bond-debt was reduced by over £10,000. Redemption and interest payments were suspended again during the parliamentary battle for the Berks & Hants Canal and, although interest was paid until 1849 in respect of the year 1844, few payments were made after this date.[143] Even if one disregards the sums expended on the repayment of loan debts incurred by compounding outstanding interest, the excess of income over expenditure never exceeded £3,000, the ratio of profit to the actual cost of building the canal (£153,000) being no more than 1½ per cent during its most successful periods before the coming of the railway.

The social and economic benefits which the canal brought to the communities and parishes through which it passed were useful but not substantial. While the canal can be said to have reduced the cost of transport, particularly of coal, and to have improved trade and agriculture in parts of Surrey and Hampshire, on balance its success was disappointingly small. It had certainly established no monopoly in the carriage of goods, since it succeeded in doing little more than provide strong competition to the waggon trade. In the Midlands and the North, when a canal had been built, industry had flocked to its banks. One might, therefore, have similarly expected that agriculture would have been substantially developed along the line of the navigation. But it was not. The tracts of sandy heath around Aldershot were still totally undeveloped when the army settled there in 1854 and the examination of a Godalming surveyor during the hearing of the railway Bill in 1834 hinted at the reason:

> Mr Wood (counsel for the promoters): Has any great extent of that Heath been chalked since that canal has been formed?
> Mr Job Smallpiece: There is not a great deal of it brought into cultivation and they cannot carry the chalk upon those commons till it is broken up.
> Wood: Is there much chalk carried upon any part of the canal?
> Smallpiece: Yes there is a great deal in the neighbourhood of Frimley.[144]

Number of Shares.	Names of Canals.	Amount of Share.		Average Cost per Share.			Price per Share.			Div. per Annum.			Dividend Payable.
		£	s.	£	s.	d.	£	s.		£	s.	d.	
1,482	Ashby-de-la-Zouch	100	0	113	0	0	80	0	0	4	0	0	Ap. Oct.
1,766	Ashton and Oldham	-	-	113	0	0	100	0	0	5	0	0	Ap. Oct.
720	Barnsley	160	0	-		-	220	0	0	10	0	0	Feb. Aug.
1,260	Basingstoke	100	0	-		-	5	0	0				
——	Ditto Bonds	100	0	-		-	-		-	-		-	April.
400	Chelmer and Blackwater	100	0	-		-	106	0	0	5	0	0	January.
1,500	Chesterfield	100	0	-		-	170	0	0	8	0	0	
500	Coventry	100	0	-		-	795	0	0	44	0	0	May, Nov.
1,851	Crinan	50	0	-		-	2	0	0				
460	Cromford	100	0	-		-	420	0	0	19	0	0	Jan. July.
4,546	Croydon	100	0	31	2	10	1	17	6				
11,810l.	Ditto Bonds	100	0	-		-	50	0	0	5	0	0	
600l.	Derby	100	0	110	0	0	130	0	0	6	0	0	Jan. July.
2,060	Dudley	100	0	-		-	52	0	0	2	15	0	Mar. Sept.
3,575	Ellesmere and Chester	133	0	133	0	0	72	0	0	3	15	0	September.
11,600	Grand Junction	100	0	-		-	243½	241½		13	0	0	Jan. July.
1,521	Grand Surrey	100	0	-		-	40	0	0	-		-	Apr. Oct.
120,000l.	Ditto Loan	-	-	-		-	97	0	0	5	0	0	Jan. July.
2,849½	Grand Union	100	0	-		-	21	0	0	1	0	0	1st Oct.
3,096	Grand Western	100	0	89	0	0 pd.	8	0	0				
749	Grantham	150	0	150	0	0	195	0	0	10	0	0	May.
	Hereford and Gloucester	100	0										
6,238	Huddersfield	100	0	57	6	6	15	10	0	0	10	0	September.
148	Ivel and Ouse Beds	100	0	100	0	0 pd.	115	10	0	5	0	0	Jan. July.
25,328	Kennet and Avon	100	0	39	18	10	25	10	0	1	5	0	September.
70	Loughborough	-	-	142	17	0	2100	0	0	180	0	0	Jan. July.
3,000	Macclesfield	100	0	100	0	0 pd.	60	0	0				
250	Melton Mowbray	100	0	-		-	200	0	0	9	0	0	July.
130	Nutbrook	109	0	-		-	-		-	6	2	0	
523	Oakham	130	0	-		-	32	0	0	2	0	0	May.
1,786	Oxford	100	0	-		-	500	0	0	32	0	0	Mar. Sept.
2,400	Peak Forest	100	0	48	0	0	65	0	0	3	0	0	June, Dec.
2,520	Portsmouth and Arundel	50	0	50	0	0	10	0	0				
21,418	Regent's	100	0	33	16	8	18	0	0	0	13	6	July.
5,669	Rochdale	100	0	85	0	0	70	0	0	4	0	0	May.
200	Stroudwater	150	0	-		-	480	0	0	23	0	0	May, Nov.
535	Swansea	100	0	-		-	200	0	0	15	0	0	November.
350	Tavistock	100	0	-		-	105	0	0				
4,805	Thames and Medway	100	0	30	4	3	4	0	0				
3,344	Ditto New	3	10	2	15	0 pd.							
——	Ditto 1st Loan	-	-	56	0	0	-		-	2	10	0	
——	Ditto 2d Loan	-	-	40	0	0	-		-	2	0	0	
——	Ditto 3d Loan	-	-	100	0	0	-		-	5	0	0	
——	Ditto 4th Loan	-	-	100	0	0	-		-	5	0	0	June.
1,150	Thames and Severn, New	-	-	-		-	30	0	0	1	10	0	June.
1,300	Ditto Original	-	-	-		-	25	0	0	1	10	0	Jan. July.
903	Wey and Arun	110	0	110	0	0	32	0	0	-		-	May.
20,000	Wilts and Berks	-	-	-		-	5	0	0	0	4	0	June.
126	Wisbeach	105	0	105	0	0	40	0	0	-		-	February.
6,000	Worcester and Birmingham	-	-	-		-	87	10	0	3	0	0	February.
800	Wyrley and Essington	125	0	-		-	115	0	0	6	0	0	February.

Table of canal share prices. In 1832 the £100 shares of the Basingstoke Canal were priced at £5, but the bonds on which interest payments had been suspended were unquoted

The railway, it was argued, would encourage farmers to do what the canal had not, but there has not in fact been any great development of agriculture in this area, partly because of the army camps established there, but equally because the land, even if broken up and manured, is not particularly suited to cultivation.

Indeed it is evident that the Basingstoke was a canal like the Portsmouth & Arundel which never justified its *raison d'être*. Not only did the promoters base their hopes on unsound trading prospects but they did not take into sufficient consideration the physical difficulties of extending the navigation beyond Basingstoke. To build a canal for 37 miles at great cost through a wild and barren countryside to a small market town in the expectation that it would be extended over a range of hills and valleys was rash speculation. As 'Wickhamensis' pointed out in 1778, it lacked the criteria for success, and even if the company had been fortunate enough to avoid its loan difficulties or to have achieved a junction with the Kennet & Avon, it is improbable that it could ever have done more than pay a meagre dividend before the coming of the railways.

CHAPTER TWELVE

SPECULATORS AT LARGE (1869–1910)

No offer to sale by tender (1869) – canal sold to William St Aubyn (1874) – revived as Surrey & Hants Canal – original company dissolved (1878) – sale by auction (1883) – attempt to form a water company – Sir Frederick Hunt – Woking, Aldershot & Basingstoke Canal Company – Hampshire Brick & Tile Company – establishment of works at Nately – temporary revival of trade – rebuilding of Frimley Aqueduct (1902) – end of commercial traffic to Basingstoke (1901) and Odiham (1904) – pleasure boating – account of second auction (1904) – William Carter – Joint-Stock Trust & Finance Corporation – Horatio Bottomley's frauds – London & South-Western Canal Company (1908).

From now on the Basingstoke Canal was to fall into the hands of a succession of speculators, few more successful than the last, who hoped with a minimum of outlay on their part to achieve what their predecessors had vainly attempted. Indeed, each of the six companies formed during the next forty years to manage the canal, was to fall into the hands of receivers. In 1869 the Master of the Rolls ordered that the waterway should be sold in one lot by private tender. The sale particulars referred to its 'extensive wharves, locks, embankments, plantations and pasture lands, occupying about 455 acres'. However, the official liquidator of the company, Frederick Whinney, arriving at his office in Old Jewry on the first day of March prepared to consider the highest offers, found that not one bid had been received. I doubt if he was much surprised, but as a consequence he remained responsible for managing the canal for the next five years.

It is probable that a section of Greywell Tunnel collapsed during this period, for Henry Taunt, the photographer and pleasure boater, writing in 1878, said that 'part of it fell in on a Sunday morning a few years since [1872], frightening the inhabitants of the village and its neighbourhood, and blocking up the navigation for several years, until the obstruction could be removed and all made good'.[145] A much later account by an unnamed observer, writing in a local magazine in the 1930s, records that 'about fifty years ago the western end collapsed only a few minutes after a boat navigated by a bargeman named Burrows, a well-known figure on the canal, and a resident of Mapledurwell, passed through'.

The remains of an iron narrow boat above lock XV at Pirbright, 1967, one of an abandoned pair, believed to be *Ada* and *Maudie*, formerly owned by Nately Brickworks. They were bought by a Richmond man who, with his wife and children, bar-hauled the pair down the canal in 1907. Delayed by a lock-stoppage, the family contracted diphtheria and had to abandon the boats, which eventually sank at their moorings

In the 1870s the development of Aldershot as a town and military centre prompted the building of further railways to link the camp with the surrounding area. In May 1870 the London & South-Western opened a line from Pirbright Junction through Aldershot to Farnham, which crossed the canal just before Ash Vale station was reached and again by the army gasworks at Aldershot, to which a spur line was laid. A branch railway built in 1879 from Frimley Junction to Ash Vale also crossed the canal by Mytchett Lake.

In July 1874 the official liquidator sold the canal for £12,000 to William St Aubyn[146] with the exception of 82 acres of adjacent land. This land was sold separately under the provision of the 1778 Act which provided that if the canal was discontinued or disused for five years the land should be reconveyed. In September 1874 an acre of the wharf at Crondall (Whites Hill) fetched £60. St Aubyn formed the Surrey & Hants Canal Company. Although very much choked up near Basingstoke the canal was, according to Taunt, in a fair way of being put into order. However, in spite of St Aubyn's attempts to revive trade, the waterway continued to lose money and in 1878 the company was put into the hands of a receiver, Mr Laulman. On 12 June a court order wound up and dissolved the original canal company. During this decade

The narrow boat *Basingstoke* about to enter Greywell tunnel, 1913

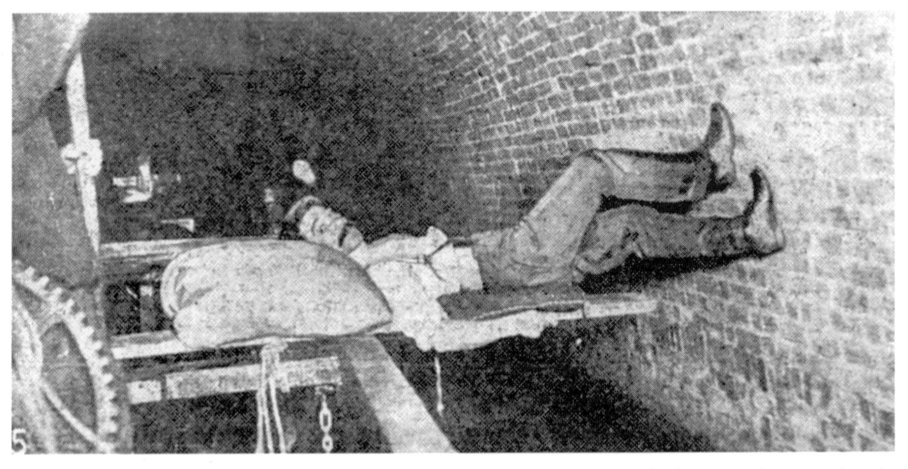

Manoeuvring a boat through Greywell Tunnel required the crew to shaft it into the tunnel and then for two people to lie on their backs on the cross plank and leg the barge through, by pushing their legs against the roof. The windlass was necessary to assist when hard aground. When Francis Giles surveyed the tunnel in 1824 he reported that its width was 15 ft at the east end but only 14 ft 1 in. at the west end

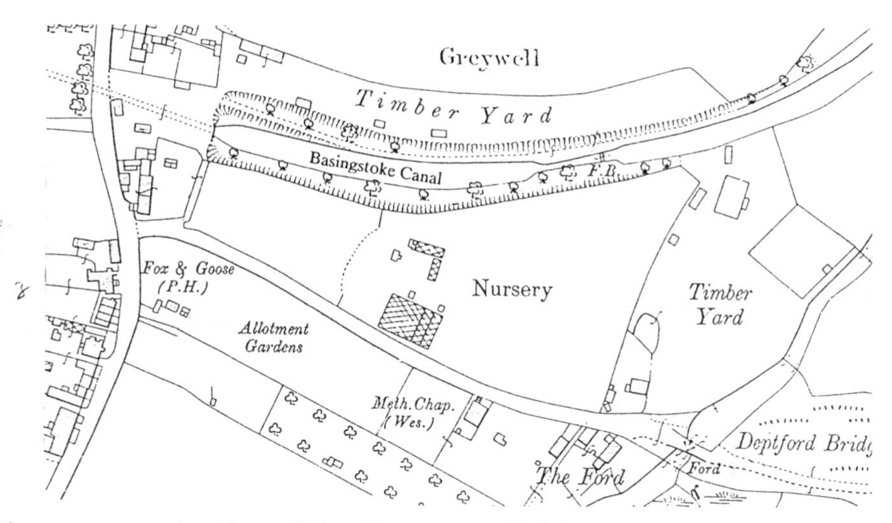

The eastern approach to Greywell Tunnel, 1909. A stop lock (lock XXX) was situated by the footbridge (F.B)

traffic to and from the Wey had dwindled every year (except two) from 8,632 tons in 1870 to 2,296 tons in 1880.

In 1880 St Aubyn sold the canal to one John Smith. St Aubyn had also disposed of some more land. A strip from Cox Heath Bridge to the edge of the Dinorben estate at Crookham fetched £32 10*s* in January 1877. Smith was a trustee for the Surrey & Hampshire Canal Corporation Limited, who acquired the land in consideration of 7,500 paid up £10 shares in the company. The authorized capital of the corporation was £100,000; subscriptions were received for 37,640 ordinary £1 shares and £68,000 raised by issuing bonds. Edward Dawson of Lancaster was the largest shareholder (14,100) and holder of bond and preference stock (£10,000). The majority of the fifty and more investors came from Lancashire and six churchmen were induced to pour £15,000 into this enterprise. Admiral Sir P.W.P. Willis of Funtingdon House, Chichester also put £4,100 into the venture. In December 1881 this company agreed to sell 4 acres of land around Basingstoke Wharf for £1,600 to meet the demands of the mortgage debenture holders. A deposit was said to have been paid to the company but this could not be accounted for and the liquidator of the company later denied that this contract was binding.[147]

In September 1880 the canal overflowed its banks and washed down part of the brick screen wall near Pirbright Junction which had been erected by the London & Southampton Railway in 1839 to reduce the risk of locomotives frightening the barge-horses. A goods train from Yeovil crashed into the rubble during the night and serious damage was caused to both engine and sixteen trucks. To what extent the Surrey & Hampshire Canal Corporation succeeded in resisting a claim for damages, is not known.[148]

The new proprietors tried to sponsor a scheme to supply the south of London with pure spring water from the canal. Reports were prepared by civil engineer William Shelford and Dr J.L.W. Thudichum* on the quality and quantity of water, but before the project could be further advanced a burst bank near the South-Western Railway at Aldershot hastened the winding-up of the corporation. Three mortgage debenture holders successfully brought an action in the Chancery Court in November 1882 to 'get in the tolls, freights and outstanding personal estate of the corporation'.[149]

The first auction was held at the Mart, Tokenhouse Yard on Friday 13 July 1883. The particulars listed rents receivable of £395 p.a., mentioned in small type that the canal was dry for ¾ mile between Ash Lock and a temporary dam at Ash Road, anticipated a considerable revenue when this break had been repaired, and offered copies of Shelford's and Thudichum's reports, of which 'the vendors do not guarantee any of the statements therein'. Several boats on the canal which had been mortgaged in May 1880 were also included in the sale.

Although no sale was effected at the auction, the three bond-holders purchased the property for £15,500 (of which £6,313 was paid in cash) and promptly formed the London & Hampshire Canal and Water Company with a capital of £50,000 in £10 shares. By 20 February 1884 £37,640 is shown as having been subscribed[150] and £88,719 was subsequently raised by debenture stock.

The canal was put back into running order and reopened in the spring of 1884. In the interim period government barges had been proceeding to Guildford from where stores had been forwarded by road to Aldershot. When trade revived, traffic showed little material improvement, the tonnage passing over the Wey Navigation rising from 301 tons in 1885, the lowest total recorded, to 3,633 in 1888. In that year the total tonnage carried amounted to only 4,187 tons, of which 3,722 were in barges owned by the London & Hampshire. Revenue from all sources totalled £1,647, a small increase on the previous year but already the London & Hampshire had passed into the receiver's hands (2 March 1887)[151] where it remained until 1895.

Apparently the attempt to market water failed. In this respect they were perhaps a little unfortunate. It was a time when water companies were being formed to provide running water and save the inconvenience and unreliability

* Dr Thudichum was also joint author of *A Treatise on the Origin, Nature and Varieties of Wine*, 1872.

In the High Court of Justice. Chancery Division. 1882. D. No. 2144.

DOUGLAS *v.* LEEMING AND OTHERS.

In the matter of the Companies Acts, 1862 *to* 1880,

AND

In the matter of the Surrey and Hampshire Canal Corporation, Limited.

Mr. JUSTICE CHITTY.

The Particulars and Conditions of Sale

OF ALL THAT

FREEHOLD CANAL & PREMISES,

LATELY KNOWN AS THE

SURREY AND HAMPSHIRE CANAL,

AND FORMERLY AS THE

Extending from Basingstoke in the County of Hants to the River Wey, in the County of Surrey, near to its Junction with the Thames, a distance of about 37½ miles.

For Sale by Auction, by

MR. SAMUEL WALKER,

OF THE FIRM OF

S. WALKER & RUNTZ,

The person appointed by Mr. Justice Chitty (the Judge to whose Court the said action and matter are attached.)

AT THE MART, TOKENHOUSE YARD, LONDON,

On FRIDAY, the 13th day of JULY, 1883, at 1 for 2 o'clock.

IN ONE LOT.

The Property may be viewed and Plans, Particulars and Conditions of Sale may be had at the office of the Canal Company at Basingstoke; at the Red Lion Hotel, Basingstoke; of Frederic George Painter, Esq. (of the Firm of Messrs. Tribe, Clarke & Co.) the Official Liquidator of the Corporation, 2, Moorgate Street Buildings, Moorgate Street, E.C., and Albion Chambers, Bristol; of Howard C. Morris, Esq., Solicitor, 2, Walbrook, E.C.; of George Legg, Esq., Solicitor, 27, Great George Street, Westminster; at the Mart; and at the Auctioneers' Offices,

22, MOORGATE STREET, LONDON.

Auction particulars, 1883.

Notice of the first auction of the canal, 1883

of wells; in the neighbourhood of the waterway, Acts had recently been obtained by Aldershot (1866 & 1879), Woking (1881) and Farnborough (1883). The canal's water supplies exceeded the needs of its commercial traffic. In 1893 agreement was reached with the Frimley & Farnborough District Water Company to allow up to 200,000 gallons to be taken from the canal in any one day on payment of one penny per 1,000 gallons but subject to a minimum payment of £150 p.a. and the free use of part of Frimley Wharf (below the aqueduct) for an underground filter-bed. The water company built a reservoir by Sturt Railway Junction but as this arrangement was repealed in the same company's Act of 1909, it may not have become effective.

The summit level continued to attract boating parties. The boathouse at Aldershot did great business at weekends during the summer months. This was especially so in 1887 when the *Aldershot & Sandhurst Military Gazette* published the account of how seven assistants from the well-established firm of uniform outfitters celebrated the public holiday granted for Queen Victoria's Golden Jubilee by rowing up the canal to Odiham and camping on the bank in their Indian tent. The *Fleet Monthly Advertiser* printed a similar account entitled 'A Trip to Odiham' but this was no more than an 'immensely enjoyable row' from Fleet. 'High Banks and almost over-reaching trees giving way to undulating meadows stretching away for miles. Leaving Winchfield Workhouse on our right . . . we found the reeds very troublesome, they having been cut down and left remaining in the water'. Beyond North Warnborough they were, however, obliged to turn back on account of the reeds.

The summer number of the *Boy's Own Paper* for 1888 published Lieutenant-Colonel Cuthall's account entitled 'Canoeing and Camping Out' of a seven-day voyage from Aldershot to Bristol. With military foresight he had had a lock key made at a blacksmith's before departing but there were so many locks 'many of them so stiff to open through want of use, that I wheeled the canoe over them in most cases'.

In 1890 the London & South-Western Railway built a steel girder viaduct over the canal above Pirbright Wharf to carry a branch line from Brookwood to serve the newly transferred National Rifle Ranges at Bisley. In 1916–17 the army constructed extensions to serve the camps at Deepcut and Blackdown. This latter service ceased in 1921 and the line to Bisley was closed in 1952. The viaduct was demolished in 1980 by British Rail.

In March 1891 Colonel Rich of the Railways Department of the Board of Trade carried out an inspection of the London & Hampshire following complaints by the rural sanitary authorities of the Hartley, Witney and Basingstoke Unions that the bridges were dangerous. His report confirmed this point. The wooden swing-bridges had rotted away or partially collapsed; the road bridges had insufficient strength to carry traction engines and heavy four-horse cart-loads of coal, which had driven out bricks from the arch faces and pushed out the parapet walls.[152] While commenting that the bridges had been well constructed but were now too light for present-day traffic, it seems unlikely that the Board of Trade was in a position to enforce his recommendations that steps should be taken to put the bridges in order.

In 1894 the Board of Trade was induced to treat the canal as a going concern with power to levy tolls, for it made a provisional order under the Railway & Canal Traffic Act 1888 which was confirmed by Act of Parliament. However, at the time when the Act was passed, the person in de facto possession was a receiver of debenture holders appointed by the Chancery Court and, as far as can be ascertained, it was not he who applied for the order but the War Office.

The waterway was purchased the following year by Sir Frederick Seager Hunt, baronet, who had been the Conservative MP for West Marylebone for ten years and was about to serve Maidstone for a further three. Perhaps because his father had been a railway contractor, the speculative possibilities of the derelict navigation attracted him and he immediately set to work to plan its revival. On 9 July 1896 he shifted the control of the London & Hampshire from the hands of its receiver by forming the Woking, Aldershot & Basingstoke Canal & Navigation Company, of which he was the major shareholder – holding 3,450 of the 3,475 paid up £10 shares.[153] The price paid by the company was given as £88,000 in the conveyance but this was juggling with figures, since he lent £20,000 to the company in 5 per cent first-mortgage debentures. There were additional loans totalling £24,950, and at least this sum would appear to have been expended in putting the canal back into order.* Some fifteen barges were built at an average cost of £250 each; others were hired. In 1897 when Taunt noted that the locks had been thoroughly repaired the total traffic had more than doubled and exceeded 10,000 tons, a considerable trade having developed in freighting Baltic timber from the London Docks to Kingston-on-Thames, Woking and Basingstoke.

There was also a considerable traffic in bricks. In the late 1880s and 1890s the wooden huts of Aldershot Camp had begun to be replaced by brick barracks. The first six months of 1896 saw no less than eighty barges from London unloading 3,625 tons of bricks at Aldershot Wharf. Bricks from London in fact accounted for half the canal's traffic that year. However, no

* The first-mortgage 5% £50 debentures, totalling £45,000, issued in August 1896, were due to be repaid on 1 July 1906 at £52 10*s* 0*d*.

THE

Woking, Aldershot, and Basingstoke
Canal and Navigation Company, Ltd.

ONE SHILLING THREE PENCE

SHARE CAPITAL, £50,000, divided into 5,000 Shares of £10 each.

No. 267 **Debenture.** **£50.**

£45,000 FIRST MORTGAGE DEBENTURES.

Repayable on the 1st of July, 1906, at £52 10s. od. for every £50, and carrying interest at the rate of 5 per cent. per annum, payable half yearly on the 1st of January and 1st of July in each year,

Issued under Authority of Article 41 of the Articles of Association and of a Resolution of the Directors.

This is to Certify that THE WOKING, ALDERSHOT AND BASINGSTOKE CANAL AND NAVIGATION COMPANY, LTD., are bound to pay to Sir Frederick Seager Hunt, Bart., M.P., of 7, Cromwell Road, South Kensington, London or other the Registered Holder for the time being hereof, the sum of £50, together with a Premium thereon of £2 10s., on the First day of July, 1906, or on such earlier day as the same may become payable in accordance with the terms of an Indenture, dated the Twenty-ninth day of July, 1896, and made between the Company, of the one part, and Albert Edward Seaton and John William Maclure, of the other part, and in the meantime to pay to him interest on the above mentioned sum at the rate and on the days above mentioned. The Registered Holder of this Debenture is entitled to the benefit of the trusts and provisions of and is Subject to the terms and conditions contained in the said Indenture.

Given under the Common Seal of the Company this 28 day of August 1896.

1900. W. No. 292. In the High Court of Justice, Chancery Division, Mr. Justice Byrne. This Debenture was produced this 3rd day of July, 1901. PAUL F. GAUNTLETT, Receiver, 15, Gracechurch St., E.C.

Directors.

Secretary.

NOTE.—No transfer of this Debenture, or of any part of the debt secured hereby, will be registered until this Debenture has been delivered at the Company's Office.

The debenture certificate issued by the Woking, Aldershot & Basingstoke Canal Company in 1896. Paul Gauntlett was the receiver when the company went into liquidation in 1900

bricks appear to have been brought from the Wey Navigation after August 1896 and this was probably due to the development of the brick-fields by Arthurs Bridge at Woking. Other efforts were made to develop trade. A letter to the Wey Navigation read as follows: 'I am asked to quote a rate for flour to Aldershot in competition with the railway company. To enable me to quote with some chance of success will you be good enough to give us a specially low bill for passing through the Wey.'[154] The reply was 3*d* a ton and consequently flour came to be carried from Battersea until 1920.

Sir Frederick also sought to bring trade to the upper reaches of the canal. Brick-making had been carried on at Up Nately on a small scale since before the canal was built and he decided to develop this industry on land bordering the canal. In October 1897 he formed the Hampshire Brick & Tile Company with a capital of £20,000 to open up brick-fields on 32 acres of woodland at Up Nately and 4 acres at Nately Scures which had been secured for £2,600. The land included a strip adjoining the north bank of the canal above Eastrop Bridge and fields both above and below Slade's Bridge. Sir Frederick and Sir William Ingram were two of the five directors and according to the articles of association none of them was entitled to remuneration until dividends equal to the capital subscribed had been paid. The initial shareholders included a Clerkenwell gas-meter maker, three city solicitors, an accountant, a cashier, a coal-merchant, a theatrical manager and Louis Simonds, the Reading brewer.[155]

The capital was fully subscribed and a cut 100 yd long, known as the Brickworks Arm, was dug above Slade's Bridge in 1898[156] to harbour the fleet of ten barges which had been procured to carry away the products of the Nately Pottery works. The canal company ledgers[157] show that the first boatload of building supplies for constructing the works reached Nately on 26 March. From then onwards hardly a day passed without the arrival of a barge laden with raw materials, and over 9,000 tons were delivered before the *Mabel* left the new cut on 13 January 1899 loaded with 15,000 bricks, of which 7,000 were bound for Mapledurwell and 8,000 for Basingstoke. This trial consignment was followed by the dispatch of 18,000 bricks in February as samples for the local builders at Odiham, Crookham, Fleet, Aldershot, North Camp, Frimley and Pirbright. By the end of the year the output of the brickworks barged from Nately exceeded two million bricks, representing half of the canal's total brick traffic which, including that of the Hampshire Brick & Tile Company, was as follows:

Year ending 30 June	*No. bricks carried on the Basingstoke Canal*	*Tons*
1896–7	489,100	1,650
1897–8	727,000	2,450
1898–9	2,133,936	7,100
1899–1900	4,034,031	13,500
July/Aug 1900	432,850	1,450

Cargoes exceeding 50 tons were rare, but occasionally up to 54 tons were recorded. The Kennet & Avon Company's *Free Trader*, drawing 3 ft 7½ in, is shown as passing on to the canal from Bristol on 5 December 1848, carrying 54¼ tons of iron.

In 1898, 20,769¾ tons produced £3,306 in tolls and freight charges; additional revenue from rents and miscellaneous items brought in a further £658 against a total expenditure of £7,041, leaving a loss of £3,077. It was claimed, however, that net revenue from all sources exceeded £5,000 in the year ending 30 June 1899 and the management at that time was contemplating success when another grave mishap disappointed the hopes of its promoters and creditors alike. The serious burst occurred by Hangdog Bridge at Woking on 15 September 1899. There was no loss of life but traffic was stopped for fourteen weeks. It was allegedly caused through mismanagement by the Woking Urban District Council in constructing a sewer beneath the canal. The council resisted the claim on the grounds that if the bed of the canal over the tunnel had been properly puddled with clay, the sewer could have been driven safely. Yet there is an interesting entry in the canal company ledgers which reads: 'Barge *Hilden* from June 25th to July 13th stationed in Hermitage Reach to prevent Mr Weldon's [the contractor] men from dredging the bed of the canal.' Because the canal had no other outlet, the disaster caused considerable consequential loss as traffic was either diverted to rail or else had to be transshipped twice from barge to cart and back. Traffic passing from the Wey Navigation to the canal fell during this period by more than half. Barges bound for places above Woking had to have their cargoes transshipped to barges above the breach or sent by land or rail. *Unity*'s load of timber for White's yard at Basingstoke was railed from Byfleet. Naturally the resultant disruption lost the carriers a lot of business. The case was heard in the High Court in July 1900, but by that time the company had resolved to go into liquidation.

The Nately Brickworks consumed some 50 tons of coal a week which was barged from Basingstoke. The bricks were freighted to many points on the canal and were used in the construction of barracks at the military training centres at Aldershot and Frimley. However, the clay was not entirely suitable, for some of the bricks soon developed faults. Sir Frederick was apparently one of the first to appreciate this, for between the end of 1899 and 23 February 1900, he disposed of his hundred £10 shares. A special resolution passed in July increased the number of directors to seven and granted them remuneration of £250 p.a. At the same time the canal company decided to charge the brick company 6*d* a ton extra toll on all traffic. Nevertheless, four weeks later the debenture holders decided to wind up the canal, by which time the Hunt family had reduced their holding to only 1,295 shares.

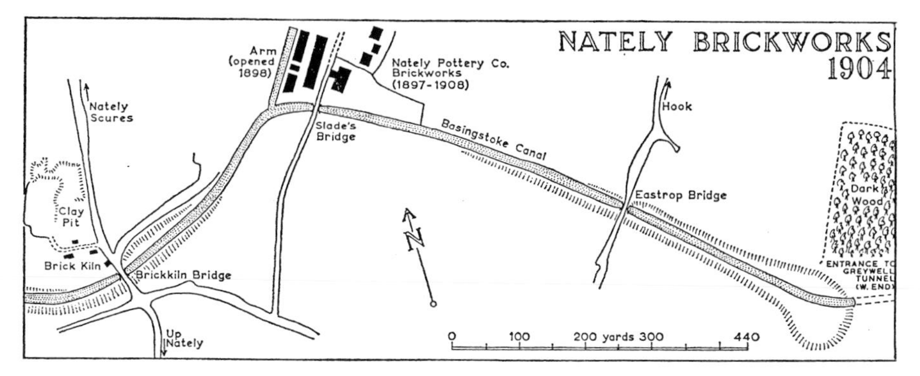

Nately Brickworks, 1904. In 1898 an arm 100 yards long was dug from the canal to harbour the fleet of ten barges procured to bring 50 tons of coal a week from Basingstoke and to carry away the products of the Hampshire Brick & Tile Company. In 1899 some 4 million bricks representing 13,500 tons were carried on the canal. After the company went into liquidation in 1901 it was renamed the Nately Pottery Company

The appointment of two new directors, one an accountant and the other a Basingstoke coal-merchant, could not retrieve the Hampshire Brick & Tile Company's position and on 8 January 1901 that company also went into liquidation. Three of the company's barges were sold to Alec Harmsworth and in the autumn the receiver invited the Wey Navigation to make an offer for the remaining seven barges.[158]

The history of the brickworks does not end there. The enterprising William Carter bought the site a few months after he had bought the canal, with a view to developing a housing estate. An elaborate prospectus was produced and 'Homesteads Ltd' (which incorporated William Carter's Estates) sought to sell sites to those seeking a home in a sylvan setting. The bricks were to be used for building the houses and it is thought that the works continued in operation until about 1908.

Besides the canal's commercial activities, increasing attention had to be paid to protecting the navigation from third parties. Not only does a moribund waterway attract more than its fair share of old bedsteads and bicycles, but items like the erection of railway bridges and footbridges, the discharge of drains, laying of gas mains, and the siting of telegraph posts, were each liable to injure the trading facilities or prejudice amenity value. In 1890 the transfer of the National Rifle Ranges from Wimbledon to Bisley caused a tramway (in use until July 1952) to be carried across the canal by a viaduct above Pirbright Bridge to the camp. The proposal by the London & South-Western Railway to double the width of its track to Basingstoke entailed the rebuilding of the

The Brickworks Arm at Up Nately. When the Hampshire Brick & Tile Company went into liquidation in 1901 and disposed of its narrow boats, the *Seagull* remained unsold and finally sank at its mooring place. In 1985 members of the Surrey & Hampshire Canal Society raised the hull and had its single cylinder inverted steam engine, built in 1890, restored

Rebuilding Frimley Aqueduct, 1902. To avoid detaining commercial water traffic for which the railway company would have been liable for heavy damages, it was decided to rebuild the aqueduct to twice the necessary width so that barge traffic would be less interrupted while work proceeded. King's Head Bridge can be seen in the background

aqueduct over the railway at Frimley. The railway company was perhaps a little unfortunate in that the timing of its decision to carry out this work coincided with one of the more successful revivals of trading, so that when it obtained its Act in August 1897,[159] the canal company had fully protected its interests. The rebuilding of the aqueduct did, of course, threaten to stop all traffic on the canal for a considerable time, so penalty clauses in the Act were certainly justified. Section 21 fixed compensation payable to the Woking, Aldershot & Basingstoke Canal Company in the event of obstruction at £20 a day and if the loss of water was so great as to reduce the depth at the overflow at Frimley Lock below 4 ft, the liquidated damages were to be £100 a day for the first six days and £200 a day thereafter until the full depth of water had been restored. Two months were to be allowed for reconstruction, with a penalty clause of £20 a day thereafter. The railway company was given the option of constructing a vertical dam to reduce the waterway by half its width for a period of two months, for the agreed sum of £1,000 and thereafter at the proportionate rate of £16 13*s* 4*d* a day. As this option did not relieve the railway company of the additional penalties for obstruction if it contracted the width by more than half – and which would in any case have slowed down building operations – the company decided to carry out the work by building an aqueduct twice the necessary width* so that traffic could be diverted into an alternative channel while the other was being rebuilt. This work was carried out in 1902, although it is recorded that in 1900 bricks from Up Nately had been used in widening the main line.[160] As far as I can trace the work of rebuilding the aqueduct proceeded smoothly and I have found no record of compensation being paid. However, it was not until July 1905 that the provision of four lines of track was completed between Clapham Junction and Basingstoke.[161]

The failure of the Hampshire Brick & Tile Company sealed the fate of the upper reaches of the canal. The carriage of corn, flour, maize, phosphates, potatoes and soda from London to Basingstoke ceased in August 1900, two months after the down traffic of timber and wheat. The barging of coal to Up Nately appears to have stopped in 1901. Empty barges continued to go up to Basingstoke Wharf occasionally but to all intents and purposes through traffic had ceased. Regular traffic to Odiham had also ceased in 1900 although the odd load of chalk continued to be brought down to Woking until 1904.

Attempts to sell the company proved fruitless, although the particulars of the canal stressed that the 'trade of the past affords no criterion as to the earning capacity of the canal per se', and went on to say rather disparagingly that it had 'hitherto been left to the management of a class of persons incapable of devising any system of organisation or of infusing any energy. Under proper management a revenue of at least £10,000 can be anticipated which would pay a dividend of 7 per cent or 8 per cent on a capital of £25,000.' The anonymous author of these particulars stated that his information had been gathered from personal visits to the canal and on the authority of persons locally acquainted with it. I do not think they were written, although they might have been, by the receiver, Paul Gauntlett, who reported in 1901 that 'the assets consist of the canal and certain barges etc, the estimated value of which is extremely uncertain'. The winding-up order, he wrote, depended entirely upon how soon the company could be sold, but it was impossible to know when this could be arranged.

> One of the loveliest canals in England. In many places it is one beautiful and ever-varying panorama of sylvan scenery framed by the rich dense foliage of arching trees, which stretch their branches quite across its bosom

* The 'double' width of the aqueduct remains evident. In 1940 stop gates were built at each end to reduce the risk of the railway being flooded in the event of its destruction by German bombing.

forming a tunnel of verdure exquisitely cool during the sultry passage of an August day, affording some of the finest camping places it has been our lot to enjoy. The towing-path is a beautiful walk, grass o'ergrown and flower-laden, the hum of bees and buzz of flies lend dreamy music to the picture of still life whilst a bevy of moorhens sport and play amongst the weeds.

Thus wrote photographer Henry Taunt in 1878. Fifteen years later the *New Oarsman's Guide* devoted two pages to the canal and described it as one of the loveliest canals in England – 'a quiet old-world waterway winding through lovely woods and heathclad moors'. However, it also noted that obstructions were numerous, the locks in bad repair and that, the section above Odiham was 'much overgrown with weeds'. 'Barges from the Thames' the guide went on 'nowadays rarely go beyond Odiham and indeed almost the only craft which ascend it now are the lighters of the War Department with stores from Woolwich to Aldershot. Cruisers are seldom seen on its waters but for those who like a quiet voyage among sylvan scrubs it is all worth a visit'. The toll was 30*s* payable at Basingstoke. There were only four lock-keepers and 'as these men have other duties, they are not always to be found when wanted; the crew must therefore carry a winch to open the locks'.[162]

There is little doubt that the popularity of pleasure boating and the prosperity of the local boat hirers greatly increased during the latter half of the nineteenth century. In 1865 licences were being issued for the use of pleasure boats on the canal and revenue that year totalled £25 from this source. On the Wey Navigation toll receipts from pleasure craft increased from £17 in 1867 to £30 in 1870 and £60 in 1874; during the early part of the twentieth century they soared to over £200 reaching £256 in 1904 and £266 in 1906.

From time to time barges would be hired out and cleaned up for a pleasure excursion. A Mr Phillips took out *Gladys* from Aldershot for a picnic outing towards Fleet on Whit Monday, 1897. On 17 August 1898 Miss Wigg of Basing took a pleasure party from Basingstoke to Greywell Tunnel aboard the *Marion* and the following summer on 22 June, Melland-Smith, the company's manager, was charged 52*s* 6*d* for hiring a barge for a similar excursion. In August 1900 the Revd Cooper-Smith took out barges *Mable* and *Enid* from Basingstoke for a church party but was only charged 50*s* a barge. Steam launches also began to make more frequent appearances; five traversed the canal in 1897 and the *Will o' the Wisp* was used for the directors' inspection in May. The first entries relating to houseboats appeared in 1898.

The auction particulars of 1883 reveal that pleasure-boat stations were run by both Henwood and Porter at Odiham, by Mrs Hill at Aldershot, and by both Belton and Bainbridge at Woking. Mrs Gregory at Basingstoke, the

Miles		
—	Basingstoke Wharf	Stats. G. W., L. S. W. Rs. Hotels, Red Lion, Wheat Sheaf. Interesting church. The first 20 miles is the summit level of the canal, and is a lockless stretch.
1¾	Old Basing House (ruins) - -	Was held by the Marquis of Winchester for the King during the Civil War. After several repulses, Cromwell stormed it on the 16th Oct., 1645. During the looting a fire broke out which reduced it to ruins.
4	MapledurwellBridge	
5	Up Nately Bridge	
5½	Greywell Tunnel (entrance) - -	Not wide enough for sculls. Is almost straight, and a boat can push through in 20 minutes.
6¼	Do. (exit)	
6¾	Odiham Castle, (ruins) - -	Once the residence of King John. David Bruce, King of Scotland, was a prisoner here for 11 years after the Battle of Neville's Cross in 1346. The canal here crosses the Whitewater, a tributary of the Loddon.
7⅛	NorthWarnborough	Post-office and public at the bridge.
8¼	Odiham - -	Hotel, George. Boat may be left at the wharf, where pleasure boats are on hire. In the George is some curious panelling from a former palace here of the Bishop of Winchester. Remains of this palace are now a farmhouse called Palace Gate.
9	Dogmersfield Park	Lovely reaches to Crookham.
13¾	Crookham - -	Inn, Chequers.
19	Aldershot Wharf -	Stat. S. E. and L. S. W. Rs. Hotels, Queen's, Imperial. The town, a mile from the canal, is uninteresting apart from its military aspect. The canal passes between the N. and S. camps.
20½	Ash Lock	
22⅝	Greatbottom Flash	} Two large ponds for storing water for the canal.
23¼	Mitchet Lake -	}
25¼	Frimley Wharf	
26½	Frimley Locks (13) begin - - -	The canal here passes down through lovely scenery.
28¾	Do. end	
29¼	Pirbright Lock -	Close by is the Guards' camp.
30½	Brookwood Lks. (3)	The lower paddles of the upper lock were immovable in 1893, but, being leaky, the lock emptied itself. On the left are the Asylum and Barracks.
32½	Goldsworth Lks. (5)	Upper lock very bad in 1893. Inn, Prince of Wales.
34½	Woking (Wheat-sheaf Bridge) -	Stat. L. S. W. R. The town is 2 miles from the canal.
36	Sherewater - -	Glorious wooded reaches.
37	Woodham Locks (6)	All in fair order in 1893.
38½	Wey Junc. (p. 11).	The distance from here to the Thames is 3 miles with 4 locks.

Extract from the *New Oarsman's Guide* (1896) regarding the state of the canal in 1893

The southern entrance to Basingstoke Wharf from Wote Street, 1904. The poster announces that the canal is once more up for auction

Revd J. Hessey and Mr Tyrell at Basing, Major-General Selby at Odiham, and Mr Steel at Aldershot each kept a boat on the canal for which they paid an annual rent to the company of a few shillings. In 1895 a list of the company's tenants showed that Belton was experimenting with a steam launch, Miss Horne of Heatherhurst Grange, Frimley, had taken up punting, Dr Fisher had given up his canoe at Crookham, both the School of Ballooning at Aldershot and the Surrey County Asylum were paying £10 p.a. for water, while the Gospel Mission was helping the company's finances to the tune of 20*s* a year for being permitted to preach on Ash Wharf. The company's rent-roll at this time was £375 p.a. but certain premises were unlet. In 1900 the annual rents receivable totalled over £700 p.a. which included £180 from timber merchant E.C. White for part of the wharf at Basingstoke, £40 from the Hampshire Brick & Tile Company for wharfage at Basingstoke, £20 from Mulford Brothers for Greywell Wharf, £25 from E.H. Manders for Crookham Wharf and £92 from Martin Wells & Company for Aldershot Wharf.

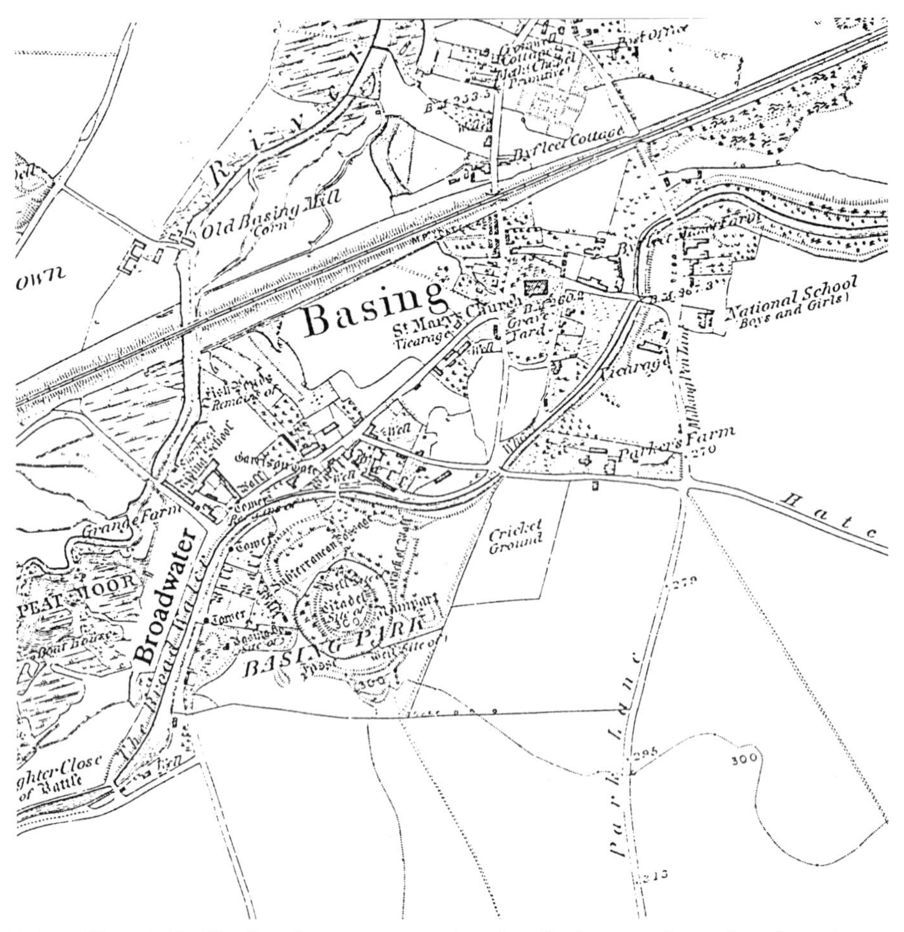

Basing village, 1873. The Broadwater was a turning place for barges and was also a favourite spot for artists and anglers until shortly before the First World War

P. Bonthron did not publish his account of his holidays on inland waterways until 1916, but it may well have been in the autumn of 1894 that he decided to 'explore' the Woking, Aldershot & Basingstoke Canal.

> Our intention had been to use our petrol launch, but this we were told would be impossible, owing to the conditions of the canal, although we were given to understand that it would be developed. We, therefore, had our skiff sent on from Henley-on-Thames to Basingstoke by rail, and on our arrival there we found the boat afloat, all ready for us. Our crew

consisted of 'Three men in a Boat', with a youth to assist generally. The first few miles down were in such a condition owing to the weeds that we found sculling impossible, and had to be towed down by horse as far as Odiham. Five and a half miles from the start we came to the Greywell Tunnel, three-quarters of a mile long, and the bargemen, in navigating their craft through, have to lie on their backs and 'tread' their way from the ceiling. The water here is as clear as crystal, the tunnel being full of natural springs. The time taken in paddling through was exactly twenty minutes, and it was a bit of work which our party appreciated thoroughly.

After stopping for tea at the George in Odiham the party sculled down to Fleet where they spent the first night at the Oaksheaf Hotel. The following day they pulled down to Aldershot and passed 'some finely-wooded parts' near Frimley Lock. From this point, their boat had to be conveyed to St John's Lock as this part of the canal was at that time out of order. Bonthron's comment was that 'from a business point of view, it struck us that if such a

Below the great wall of Basing House with two of its octagonal dovecotes built, *c.* 1530. Notice of the forthcoming canal auction to be held in October 1904 can be seen displayed on the fence on the left

canal as this could only be deepened it would make a fine through waterway from London, via the Thames'.

With the failure of the Woking, Aldershot & Basingstoke Canal Company, traffic on the canal dwindled to a mere thousand or so tons. In January 1904 Sir Frederick Hunt died, and Mr Justice Warrington agreed on 5 February in connection with the action by *Ingram* v. *the Company and Others* that the property should be sold. The auction was held in London at Tokenhouse Yard on 27 October 1904.

An article entitled 'A Canal for Sail'[163] described the scene in Room A at the Auction Mart. The writer recounted that he had been encouraged to attend the sale by the fact that he once knew

> . . . a member of the L. & S.W.R. Co's clerical staff who left the service to become General Manager of the Basingstoke Canal Company [Mr Melland-Smith was manager and engineer] for the munificent salary of 30s per week, and a horse and trap to convey him over the ground. He subsequently deplored the step and averred that the trap was more in evidence than the horse. Here was both the opportunity and the inducement to see for ourselves what is understood to be the first sale by auction of an English Canal.* Punctuality is not 'the soul of business' at Auction Marts, and whilst waiting for the arrival of the auctioneer, we were left with no better occupation than that of scanning and estimating the object if not the capabilities of the other onlookers, as the tailors say 'taking their measure', and a more motley assemblage, consisting of about 160 persons, it would be difficult to imagine. There was the inevitable sprinkling of lawyers who had some connection with the Canal in the past or hoped to have the pleasure (and the profit) of doing so in the future; a few men who were apparently bent upon getting something 'at a sacrifice', a few more who, like ourselves, had come to look on and take notes, and the remainder of heterogeneous collection of humanity who had 'dropped in', either from idle curiosity or for a temporary escape from the dense fog which hung like a pall over London and lent additional gloom to the already dismal surroundings.
>
> At last the auctioneer, Mr Benjamin I'anson Breach, appeared, and having taken his place on the rostrum with due ceremony combined with an impressive air, opened the proceedings with an elaborate and glowing word-picture of the property and its condition in his blandest tones and

* The canal had, of course, been previously put up for auction in 1883, as had the Wey & Arun Junction Canal in 1870.

In the High Court of Justice.

1900.—W.—No. 2922.

CHANCERY DIVISION.
Mr. JUSTICE WARRINGTON.

Re The Woking, Aldershot and Basingstoke Canal and Navigation Company, Limited.

INGRAM—v.—THE WOKING, ALDERSHOT AND BASINGSTOKE CANAL AND NAVIGATION COMPANY, LIMITED, AND OTHERS.

HAMPSHIRE AND SURREY.

To Canal and Railway Companies, Carriers, Government Contractors, Riparian Owners and Others.

Particulars, Plan, and Conditions of Sale

OF THE

FREEHOLD

(AND PART COPYHOLD AND LEASEHOLD)

COMMERCIAL PROPERTY

KNOWN AS THE

WOKING, ALDERSHOT AND BASINGSTOKE CANAL

Commencing in the Town of Basingstoke, having a length of

Thirty-seven Miles

Through some of the most picturesque and residential neighbourhoods of Hampshire and Surrey, and terminating at the junction of the River Wey, by means of which a **direct line of navigation is opened to London**, a distance of nearly seventy miles. Together with all

LOCKS, WAREHOUSES, COTTAGES, LANDS AND WHARVES;

the whole embracing a total area of about

384 ACRES,

Lying in the Parishes of Basingstoke, Eastrop, Basing, Maperderwell, Up Nately, Greywell, Odiham, Winchfield, Dogmersfield, Crookham, Fleet, Hawley with Minley, Cove, Aldershot, Ash and Normandy, Frimley, Pirbright, Woking, Horsell and Chertsey. The whole forming

A GOING CONCERN, WITH POSSESSION,

Which will be Sold by Auction by

Mr. B. I'ANSON BREACH,

OF THE FIRM OF MESSRS.

FAREBROTHER, ELLIS, EGERTON, BREACH & CO.,

With the approbation of His Lordship Mr. Justice Warrington, the Judge to whom the said Action is attached, pursuant to the Order therein, dated the 5th February, 1904,

AT THE AUCTION MART, TOKENHOUSE YARD, CITY OF LONDON,

ON THURSDAY, OCTOBER 27th, 1904,

At TWO o'clock precisely. **In One Lot.**

Particulars, with Plan and Conditions of Sale, may be obtained of Messrs. MADDISON, STIRLING & HUMM, Solicitors, 6, Old Jewry, London, E.C.; Messrs. ROLLIT & SONS & BURROUGHS, Solicitors, 3, Mincing Lane, London, E.C.; Messrs. NICHOLSON, GRAHAM & GRAHAM, Solicitors, 24, Coleman Street, London, E.C.: Messrs. ELAND, NETTLESHIP & BUTT, Solicitors, of 4, Trafalgar Square, London, W.C.; Messrs. GOLDING & HARGROVE, 99, Cannon Street, London, E.C.; Messrs. CHURCH, PRIOR & ADAMS, Solicitors, 11, Bedford Row, London, W.C.; of the Receiver and Manager, PAUL E. GAUNTLETT, Esq., Chartered Accountant, 6, Rood Lane, Fenchurch Street, London, E.C.; at the Auction Mart; and of

Messrs. FAREBROTHER, ELLIS, EGERTON, BREACH & CO.,
29, Fleet Street, London, E.C., and at the place of Sale.

NOTE.—*For Order of Sale see "The Times" or "The Standard" of the day preceding the Auction.*

H. S. CARTWRIGHT, Printer, 19, Southampton Buildings, W.C.

Notice of the second auction in 1904. Sales of canal property had reduced the total area by 71 acres between 1869 and 1904. There were no bidders at the auction and the canal was sold by tender to William Carter in 1905 with such fervour as only an auctioneer of the highest standing can command or find adequate language to portray. He was 'present by the command of the Court of Chancery to dispose of the lands, wharves, warehouses, cottages etc. the whole embracing an area of rather over 366 acres'. Proceeding, the auctioneer said the whole of the locks were strong and well constructed, and in good working order. We hope we shall be pardoned if we misjudged their motive, but we thought we detected an uncomfortable movement towards the exit on the part of two or three of the audience who, upon the bare mention of strong locks in good working order, sought the protection of the prevailing fog.

Mr Breach said he hoped the purchaser (for of course the financiers present would tumble over each other in their struggles to secure such an El Dorado) would do nothing to mar its unspeakable beauty in his efforts to raise a great profit, at which declaration and mention of 'profit' a titter ran round the room but, affecting not to have observed it, the auctioneer hastened to assure the audience that it had once been a good paying concern but that was before the wicked South-Western Railway ran

Canal Cottage, Chequers Bridge, Crookham, stands south of the canal to the east of the bridge on the road between the village and Crondall Cross roads. This scene was taken at the time of the 1904 auction and shows the canal company's maintenance boat marked as lot 124

through the district, but still there was no reason whatever why it should not pay again – even a greater dividend than before (another titter) as it had produced as recently as 1899 an income of £5,000, and he drew a vivid picture of what might be done with sailing yachts or motor power. If his audience would accept his word and an enterprising purchaser would obtain an Act of Parliament and extend the Canal to Southampton, it would become an invaluable asset and a splendid highway from that port to London. These remarks were greeted by some cynics in the audience with congratulations that some part of the Canal was in close proximity to Brookwood Asylum, and when the auctioneer announced that there was 'a large sum of money in the Basingstoke Canal' they could restrain themselves no longer and the proceedings were interrupted for some minutes until the laughter had ceased – the mirth being prolonged by someone announcing that he had 'made up his mind to sail away'. Then these sceptical millionaires having vented their hilarity, put some pertinent questions to the representative of the vendors, and some who 'knew something' ventured to assert that the Canal only existed in very wet weather when the clouds supply the water way, and long sections of the nominal Canal were dried up, but this was said to be due to draining the water off to admit repairs being effected. Great Scott!! We lived for two and a half years in a house the garden of which bordered the Canal path, and have frequently walked across it when no repairs were being effected. In answer to another inquisitive person the auctioneer had to admit that it would be necessary to get a short Act of Parliament passed before sailing yachts or motor boats could ply for hire. Another roar of laughter greeted the announcement that there would be no opposition to these Acts.

At last the auctioneer, foreshadowing an 'all-night sitting of the house', suggested in his sweetest tones that the would-be-purchasers were rambling from the subject. 'You are getting into Acts of Parliament but I want to get into the Canal.' 'And I hope you'll stop there,' retorted someone with a rubicund countenance having the appearance of a farmer. Mr Breach's patience appeared to be waning. 'Now gentlemen, to come to the point. What will you give for the Basingstoke Canal?' For the first time, a dead silence ensued and everybody looked at everybody else as if anxious to catch sight of the living curiosity who wished to become the possessor of this premier highway from Southampton to the Metropolis. 'Shall I say £50,000 for the 366 acres, wharves, warehouses and cottages?' No answer. 'May I say £40,000? £30,000? Gentlemen; am I to go back to the Court of Chancery and tell them that nobody wants the Basingstoke Canal? Will nobody make me an offer of only £20,000 for it?'

But no answer came to the gentleman's pleadings. 'Very well gentlemen, I can sell it privately for more than £20,000 without any trouble,' remarked the auctioneer, but what caused this outburst of generosity in offering it for a lower sum to an unappreciative and ungrateful public was not so apparent as the contempt of the audience!

In the following year the canal was put up for sale by tender although this time the area had been reduced by 18 acres (of which 16 were water) which had been purchased by the War Department.* It was at this stage in the canal's history that financier Horatio Bottomley, Liberal MP for South Hackney and founder of *John Bull*, entered the scene and the waterway became the subject of a gigantic swindle. In August 1905 the receiver had sold it to William Carter, a pottery manufacturer, of Parkstone, for £10,000. *The Times* commented on 18 August that the purchaser 'will probably endeavour to make the canal pay with a service of motor boats, and the next most practical suggestion which has commended itself to him is to run off the water and turn the course into the trunk line of a light railway with tramway extention in the most populous centres.'†

Carter, having apparently bought the canal 'on spec' and not being quite sure what to do with it, sought out an old acquaintance, the fraudulent Ernest Hooley at his Bond Street office. Felstead tells the story.[164]

'Hooley,' exclaimed the new-comer. 'I've bought a canal!'

'Good Lord,' cried Hooley, 'what for? What are you going to do with it?'

'I don't know; it only cost £10,000, so it must be cheap. It's the Basingstoke Canal and it's thirty-three miles long. I've bought everything – wharves, boats, cottages, freehold, everything.'

Hooley groaned. 'My dear Mr. Carter,' he said, shaking his head ominously, 'do you know this infernal canal has already broken half a dozen men?'

'What?' with great incredulousness.

'It's a fact; you must have been crazy.'

Carter's face dropped perceptibly; the brightness went out of it and it was

* The War Department purchased the freehold of Great Bottom, Eelmoor, Puckridge, Claycart and Rushmoor Flashes in 1904.

† An article in the *Motor Boat* referred to 'The temptation to apply the most modern invention to the most out-of-date object, with the idea of resuscitating that which had already served its purpose, seems to be irresistible'. Consequently all sort of strange proposals had been suggested in connection with the waterway including one that it should be used for motor-boat races and speed trials. The writer had met the new owner, Mr Carter, the Chairman of Nately Pottery Co., considered him 'energetic', and reported that he hoped to develop pleasure boating and introduce a twice-daily passenger boat service between Aldershot and Basingstoke. However, the writer concluded, 'for the present, communication with the Wey and the Thames is out of the question' as the constant passing of water would probably drain the upper part of the canal!

in a much more humble tone of voice that he asked if Hooley couldn't do something to get him out of his trouble. 'Can't you float it?' he inquired.

'Float it?' retorted Hooley. 'There isn't enough water in it.' (He always would have his little joke.) 'Anyhow, leave the matter to me and I'll see what I can do.'

Hooley made enquiries and found out that the canal had been keeping the solicitor (who was also the receiver) and his family for twenty years. 'The best thing I can do for you is to take you along to see Bottomley,' said Hooley. 'He'll be able to relieve you of it if anyone in the world can' – which was perfectly true.

Max Beerbohm's cartoon of Horatio Bottomley, 1912. Bottomley stands looking pensive with several writs stuffed into his pocket, encircled by a crowd of lawyers

Broadwater, Basing village, 1904. Barges had by now ceased to go regularly to Basingstoke

Bottomley had formed in 1904 the Joint-Stock Trust & Finance Corporation with a capital of £12,000 which had been subsequently increased to £500,000 in 5*s* shares. Carter agreed to sell the undertaking to the Joint-Stock Trust for £250,000 in shares of that company. They in turn resold it to Hooley for £150,000 on the understanding that he formed a company to repurchase the canal for £200,000. A company was registered for this purpose but did not acquire the property. Subsequently it was to be revealed that striking irregularities had occurred, including the duplication of shares on an enormous scale. Bottomley (for the second time) was prosecuted for conspiracy to defraud and the case was heard at the Guildhall over a period of three and a half months, at the end of which, in February 1909, Bottomley was surprisingly acquitted without the case even going to the jury.

In November 1907 the liquidator of the Joint-Stock Trust sold the interest, if any, of the Trust to Hooley's nominee, Mr J.A. Campbell, for £500, the sale being approved by the court, and in February 1908 he formed the London & South-Western Canal Company with a capital of £100,000 (£68,000 in fully paid shares, £20,000 in debentures and £12,000 by mortgage to Carter). It was obvious that Bottomley hoped the title might lead people to think that it was connected with the London & South-Western Railway and both he and Hooley sold shares to a number of personally selected victims. The company registers show that the 10,000 shares held by the John Bull Investment Trust Agency (controlled by Bottomley) and the 68,000 by Campbell were disposed of within a year. Some were bought by Holroyd, the Byfleet brewer, who became a director of the company. Two young men, Reginald and Vincent Eyre, purchased over 55,000 shares.[165] Reginald was only twenty-five when he had met Bottomley at a racecourse and his brother, a lieutenant in the 1st Life Guards, only twenty-one. When the gullible brothers discovered they had exchanged their gilt-edged securities for worthless stock, they brought an action against Bottomley in July 1908, alleging conspiracy and fraud. Bottomley took immediate action. Alarmed at the prospect of being found guilty of such charges, he had Vincent Eyre tracked to Paris and there, in company with his faithful retainer, Locke Cox, familiarly known as 'Tommy', he obtained from the boy, during a sumptuous dinner, a letter withdrawing all the allegations. Bottomley then consented to judgment against himself for £16,000. Next year the other Eyre brother sued Bottomley, and got a verdict by consent for £28,000, but whether either of the boys ever got all their money is doubtful. Bottomley was always careful to retain some loophole through which he could subsequently wriggle in any law case. When he was asked in relation to the Eyre case why he had released the brothers from liability if the transactions were honourable, he replied: 'Because they were not liabilities of mine. I had nothing to do with the transactions. You had better ask Mr Hooley. He settled the action and paid every penny necessary.'[166]

Horatio Bottomley was a scoundrel, but even some of those who suffered from his fraudulent wiles, could not but admire the charm of the 'rogue hero'. On one occasion he was confronted in court by Montagu Lush KC. Lush was an unimposing figure, short, squat and afflicted with a high treble voice. Bottomley on the other hand appeared calm, deliberate, frigidly impassive, his voice firm and resonant, betraying neither anger nor trepidation. In the course of his cross-examination Bottomley mentioned the flotation of the London & South-Western Canal Company.

'Oh,' squeaked Lush, 'is that our old friend the Basingstoke Canal?' Horatio eyed his interrogator with an almost pitying look.

'I don't know anything about your old friends, Mr Lush!'

Titters of laughter, then the cry of silence in court. Mr Lush made no further mention of his "old friend". Not until 1922 was Bottomley convicted and sent to gaol on similar charges and the Joint-Stock Trust was not finally dissolved until after his death in 1933.

In July 1909 Carter gave notice requiring repayment of his mortgage and in November the London & South-Western Canal was wound up on a petition of Reginald Eyre.[167] When the accounts were issued, liabilities totalled £35,911, of which £13,263 represented Carter's mortgage and interest and £22,648 ranked against assets of £27 to meet the claims of debenture holders. A deficiency of £91,660 was shown as regards contributories. Proceedings were instituted against certain of the directors to recover the amount of their qualifying shares which they had received as gifts from the vendor, but although judgment was given in favour of the receiver, he concluded the company's dismal record by announcing in August 1910 that it had not been found possible to recover enough to discharge the costs of the proceedings. There were many who had lost money through the machinations of both financiers, and to this day mention of the canal brings painful memories to the families of those who believed the plausible tales of Bottomley and Hooley.

CHAPTER THIRTEEN

THE HIGH COURTS OF JUSTICE (1910–13)

Nature on old canals – the bridges of Woking – Woking UDC (Basingstoke Canal) Act (1911) – position of William Carter and the London & South-Western Canal Company – High Court decides company is liable to pay certain costs – Court of Appeal allows appeal (1913) – an anomalous situation.

By the turn of the century naturalists were starting to draw attention to our neglected waterways as sources of abundant wildlife and on 1 January 1910 *The Times* published a lengthy and typical nature article on old canals. Its author was clearly familiar with the Basingstoke Navigation where it ran 'through the thirsty plains of Aldershot' and commented on how the banks of the canal were brightened in summer by troops of gem-like dragonflies.

> A few of the larger species, familiar in lanes and gardens, sweep from the bank and circle in summy clearings among the fir trees with an audible rustle of their wings; but they are far outnumbered by the smaller and more brilliant kinds that seldom travel far from the waterside. One beautiful species, with large wings blotched with deep blue or dull red, according to sex, seems perpetually drifting from one green sedge stem to another, and never leaves the shore. But flights of another and smaller kind, with a brilliant sky-blue body, wander at times some distance up the heathery slopes; and when a dozen or twenty of these brilliant little creatures leap suddenly into sparkling flight above the purple heather-bloom it is a sight of fairylike splendour.
>
> The old locks on deserted canals have provided for many years past a favourite resting place for many birds and flowers. Where the surface of the bricks has scaled away and the mortar has perished from between them the sprouting seeds take hold, and the motionless water of the lock reflects the hanging blossoms. Dwarf bushes of hawthorn and elder fleck with flower the mellow red of the wall; and small climbing plants like the wild strawberry and 'mother of thousands' fling their delicate strands from root-hold to root-hold upon the brick.

However attractive the canal had become to some, to the Woking Urban District Council it was a source of great vexation, and as a result of their

Two narrow boats, *Ada* and *Maudie*, owned by the Hampshire Brick & Tile Company, seen at Little Tunnel Bridge, *c.* 1900

determination to safeguard the public and the ratepayers from the perils and expense of a derelict waterway, a legal battle was about to commence which was to have unexpected consequences.*

In 1910 the council had circularized local authorities through whose district the canal passed, drawing attention to the dangerous condition of many of the bridges and the need for new ones, and stating that from the history of the canal's ownership it was likely to be a long time before any person or company in a sound financial position would acquire the property. Local authorities were invited to say whether they would be prepared to join the council in applying for an Act of Parliament to authorize them to rebuild or repair the bridges. At a conference held in Woking Town Hall in April to discuss the canal's future, Carter claimed only £4,000 was needed to render the canal navigable throughout and offered to sell the canal to the local authorities for £15,000 payable in 4 per cent loan stock. Carter's proposals were unacceptable (the terms and conditions of sale were never defined) and

* In 1906 the Royal Commission on Canals and Waterways heard evidence about the condition and prospects for inland navigation in the British Isles. One witness stated that no barge had run from Aldershot to Basingstoke for 'several years now', but that it would be well to revive the canal; he also stated that Surrey County Council were waiting for a wealthy purchaser to rebuild the bridges!

in November Carter's solicitors announced his intention of applying to the Board of Trade for authority to close the canal.

The Woking UDC, however, had decided meanwhile to go ahead and obtain an Act to enable them to repair and widen the six road bridges (Monument, Guildford, Wheatsheaf, Arthurs, Hermitage, Brookwood) and one footbridge (on Horsell Moor) in their district since they alleged the canal company had 'for many years made default and have persistently neglected to support maintain and keep in sufficient repair at their own expense the canal bridges notwithstanding the representations and complaints of the council that the same could not be used for traffic without risk of serious danger to the public'.[168] The council said the canal company was insolvent, the undertaking derelict, the navigation disused and that they were unable to recover expenditure of £443 already expended on repairs to five bridges. In its passage

Three bridges crossed the canal in Basing village. This 1900 view is of Church Lane Bridge, which carried the lane from Milking Pen Lane to the village centre. The vicarage is beyond the bridge to the left and Basing Wharf (already termed 'old' on the 1910 map) is beyond on the right

through both Houses the Bill was amended to ensure that the width of the waterway was not reduced to less than 14 ft 9 in., nor the headway above the tow-path to less than 6 ft 6 in. and the notice of intention to begin work (except in urgency) was extended from at least eighteen hours to seven days. The Act granted on 18 August 1911 empowered them to reconstruct the bridges and to recover the cost from 'the canal undertaking'. It also contained sections to protect the Basingstoke Canal Navigation Company (Section 11) and compensate the Woking District Gas Company (Section 34) in the event of its coal supply being interrupted by the intended works. The council did not have as good an Act as it had hoped, but it was, the council thought, good enough.

On 7 November *The Times* printed a plea from a Mr Southgate Day to save the waterway as negotiations were in progress to throw the bridges into the channel and to drive roads across the bed. His arguments for preservation can be echoed today – 'In the summer its waters are gay with boating and picnicking parties and the angler fishes while in the winter there is skating. There are also delightful walks along the leafy tow-path.'

The following year Woking UDC, seeking recompense for the work they had carried out and realizing that any action against the defunct and amoebaean London & South-Western Canal Company would be a waste of time, sued William Carter as mortgagee in respect of the expenses paid and incurred by the council. The case was set down at the Woking Petty Sessions and adjourned at the first hearing on the question of the mortgagee's liability. Before it was resumed, counsel for both parties tried to reach agreement, but, as was reported in *The Times* on 19 February 1912, 'it was found that although the council were acting upon the advice of Mr Moresby and Mr Carter was acting upon the advice of the conveyancing counsel, there was an absolute conflict between the opinions expressed by counsel, each thinking that their clients were absolutely in the right'. In the circumstances it was then considered to both parties' advantage to reach a compromise and it was agreed that the council should have a charge upon that part of the canal within their district for half the sum claimed, which would have priority over Carter's mortgage. The proceedings were thereupon withdrawn.[169]

The position remained unsatisfactory for both parties and since the London & South-Western found themselves designated by the Woking Act as possible successors and assigns of the original company of proprietors formed under the original Act, they took out, with their mortgagee Carter, a summons asking whether according to the true construction of the Act of 1911 they were either of them liable for any, or for what part, of the costs incurred by the Woking UDC pursuant to the Act. Although the amount in question was not large, the question of liability for the repair of the bridges was a matter of serious importance.

The swing-bridge at Basing was replaced by a fixed wooden bridge, *c.* 1910. This view shows the road, left, which led from Basing past Slaughter Close to Upper Mill, whose outbuildings can be seen beyond the cart. Swing Bridge Cottages were built in the 1840s and are still inhabited, but the wooden bridge collapsed and the bed was filled in during the 1950s

In the High Court Mr Justice Sargant held that the London & South-Western were liable as the proprietors of the canal for any costs incurred by the council. He further held that the property was liable to a charge in respect of such costs, both in the hands of the canal company and William Carter. The company appealed and briefed Mr Martelli KC and Mr C.E.O. Carter to represent them. Mr Romer KC and Mr A. Underhill appeared for the respondent council. The case was heard by the Master of the Rolls, Lord Justice Swinfen Eady and Lord Justice Phillimore. At the conclusion of the arguments their lordships adjourned to consider their decision and in the meantime Carter asked the local authorities to join him in making it a workable concern and stated that £4,000 would put the canal in order.[170]

It was not until Friday 28 November 1913 that the Court of Appeal of the Supreme Court of Judicature gave judgment in the matter of the Woking Urban District Council (Basingstoke Canal) Act, 1911. All three judges unanimously agreed that, while the circumstances of the case were peculiar and the question of law difficult, neither to the London & South-Western Canal Company nor to Carter the mortgagee nor upon the property could liability be attached. Mr Justice Sargant's view that the Act of 1911 imposed any fresh liability could not be accepted. The success of the appeal rested on the interpretation of the words 'their successors and assigns' which were held to be meaningless as words of limitation in a conveyance to a corporation aggregate. In the case of a statutory company formed to carry on a public undertaking it was clear that the company could not assign the undertaking with its rights and obligations. The judgment of Lord Cairns in Gardner's Case applied, for there was for this purpose no distinction between a canal company and a railway company. The words 'their successors and assigns' were meaningless and must be disregarded.

In this respect it was necessary to consider the precise operation and effect of the conveyance to St Aubyn in 1874. The statutory undertaking still

Upper Mill, Basing, 1904. It was shown as a cornmill on the 1872 map and as Upper Mill Farm in the 1930 revision. The farmhouse was pulled down in the 1950s but the dilapidated farm buildings remain

remained in the company. Even if, contrary to his lordship's opinion, the conveyance passed the canal to the purchaser, the rights of the public to use the canal could not be destroyed by the conveyance. The purchaser could not build over or destroy the canal, or legally levy any tolls. On the other hand, he was not liable to the obligations to repair bridges etc. There were some observations by Lord Justice James in *Re* Bradford Navigation Company which suggested that the purchaser might take the land subject to the rights of the public. He (the Master of the Rolls) could not regard those observations as justifying the inference which it was sought to draw from them. There was also a dictum by Lord Lindley in *Marshall* v. *South Staffordshire Tramways Company* (1895) to the same effect. But his lordship was unable to assent to that view. In his opinion nothing passed by the conveyance, for it purported to pass that without which the statutory undertaking could not be carried on, and was therefore *ultra vires* the company. In saying that, he did not refer to any surplus lands, or to property which might be parted with, without damage to the undertaking. The proper course for the liquidator to have taken was to have applied for a private Act of Parliament authorizing the transfer of the undertaking to the purchaser. That course had actually been followed in several cases. In his lordship's opinion the purchaser acquired only a possessory title which might ripen into a fee simply by virtue of the Statute of Limitations.

Then in 1878 the company was dissolved by an order of the court. No such order ought to have been made, but it was made. The effect of the dissolution of a corporation was that its lands and tenements reverted to the person or his heirs who granted them to the corporation. A right of entry therefore arose in 1878 in favour of persons claiming through or under the persons who conveyed to the company shortly after 1778. That right of entry had, of course, been barred by the Statute of Limitations. The result appeared to be that St Aubyn and persons claiming through or under him had acquired the legal fee simple in the canal, including the locks and wharves, free from any obligations imposed upon the original company and without any of the rights conferred upon the company. There had been several conveyances since 1874 to persons deriving title through St Aubyn. The last purchaser was the London & South-Western Canal Company, who executed a mortgage in favour of Carter. The company was in liquidation, and Carter was mortgagee in possession. The position of the limited company was that it was owner of the waterway, the canal – it was not bound to effect any repairs or to keep up the canal, and any payment of an agreed sum was not a toll and could not be claimed as such. In 1894 an Act was passed confirming a Provisional Order for (*inter alia*) 'the Basingstoke Canal Navigation'. There was no ground for holding that this Act by implication created a new corporation. Even if this were so, it would not avail the present limited company, whose title was only acquired in 1908 and who could not

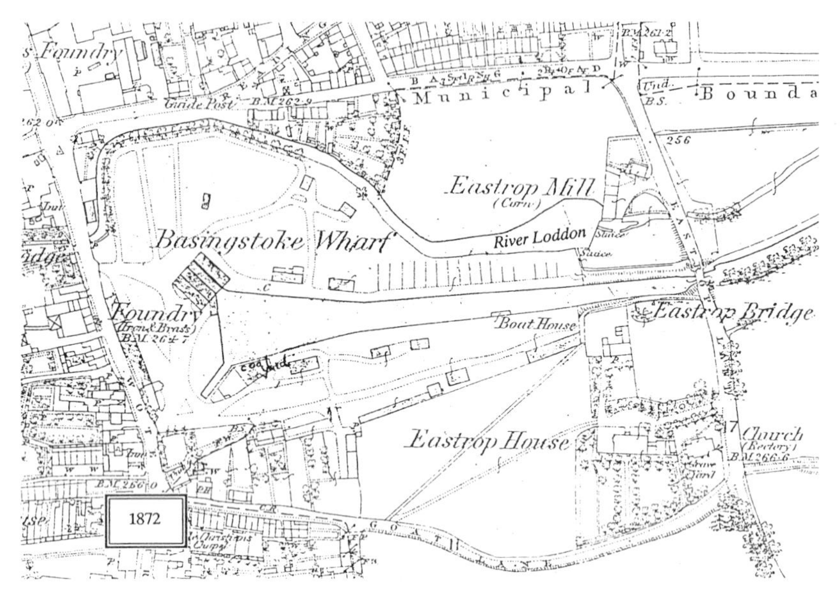

Basingstoke Wharf, 1872. It was described in 1869 as containing the wharfinger's residence, outbuildings and large garden; three brick- and slate-built roomy offices; large corn, coal, bark, deal and timber sheds and storehouses; sawpits and sheds; two powerful cranes and other conveniences for loading barges; extensive timber wharf; gatekeeper's residence and garden; six thatched cottages

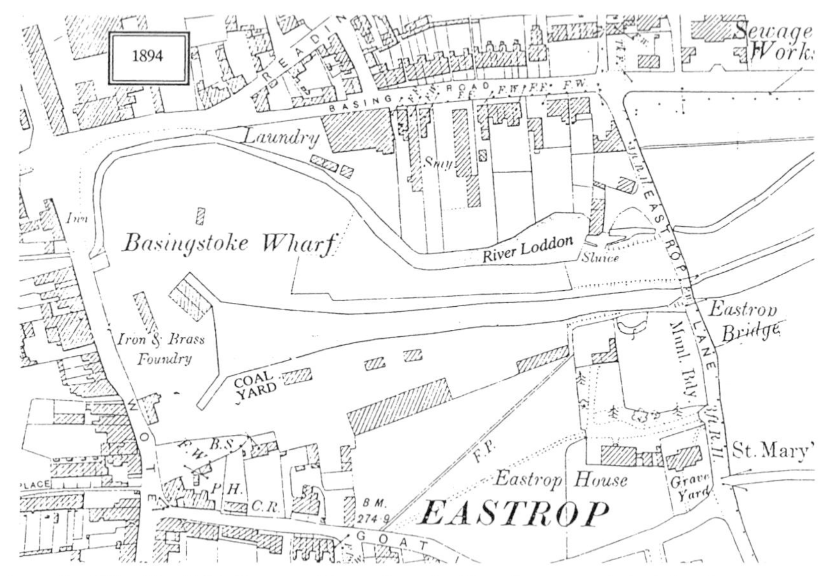

Basingstoke Wharf in 1894. Barges still use the wharf but the covered dock has been dismantled. The boat house has also vanished

The only known photograph of the canal basin at Basingstoke. This 1904 view shows two barges belonging to the Hampshire Brick & Tile Company which had been used to convey coal to the Nately Brickworks, opened in 1898. The company went into liquidation in 1901 and seven of its ten barges were sold to the Wey Navigation

obtain by purchase the undertaking, which, *ex hypothesi*, was vested by statute in some corporation, the name of which was not defined.

The effect of this decision was that the Act of 1911 did not impose any fresh liability. It authorized the council to advance the money and to do the work, and to recover a proper proportion from any person liable, whether *ratione tenurae* or otherwise. 'It was obviously a matter of doubt and difficulty. But if neither the limited company nor Carter fell under the head of "the company, their successors or assigns", or could claim to own the undertaking, no liability could attach upon either of them or upon the property.'[171]

As a firm of solicitors commented: 'Mr Mossop has been placed in the position of Napoleon at Waterloo.'[172] A week later Mossop, the clerk to the Woking UDC, took pains to emphasize that they had been justified in seeking their Act and that the court's decision was based entirely upon points of law, no points being brought out which were not known to the council when they applied for the Act. The effect of the Court of Appeal's decision was that although the London & South-Western Canal Company had obtained a title through lapse of time, they had none of the rights to make by-laws or levy

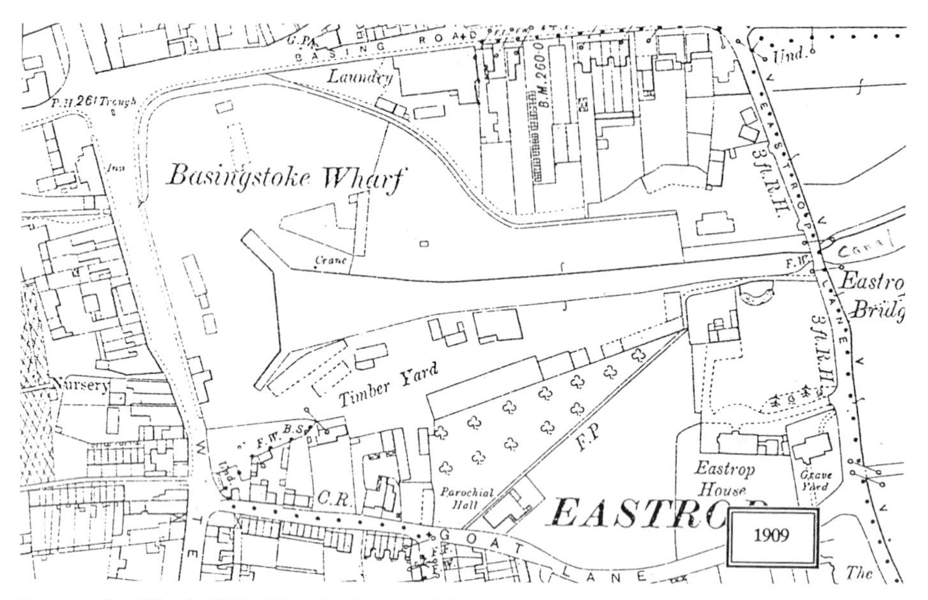

Basingstoke Wharf, 1909. The wharf covered 6 acres. It had an extensive frontage to Wote Street and also to Goat Lane. The group of buildings comprised the two-storey brick office with slate roof, a large shed with four bays overhanging the north-western arm, a range of timber and thatched sheds, stabling, a large crane, a timber yard with saw pits and the wharfinger's residence which had ten rooms as well as large fruit and kitchen gardens. In 1904 it was stated that there was 'a splendid supply of water flowing from the well all day and night' which was pumped through 15 in. pipes from a well to the basin

tolls conferred upon their original company, nor were they under any liability to repair the bridges, both rights and liabilities having ceased to exist when the first company of proprietors was dissolved in 1878. If the owners of the canal were not liable to repair the bridges, the point arose as to who was liable; and it appeared that, under the common law, all bridges erected prior to 1803 outside cities and boroughs, if of public utility, must be repaired by the county unless by usage, prescription or tenure the duty to repair fell upon someone else. The onus was upon the county to show that a bridge erected before 1803, and of public utility, was not a bridge reparable by them. As the canal was opened for traffic in 1794, and the company had had to erect bridges before cutting through any highway, there was no doubt that these were built previous to 1803. In these circumstances it appeared to Mossop that there were substantial grounds to support the contention that the liability to repair bridges devolved upon the county council, as since 1878 there had

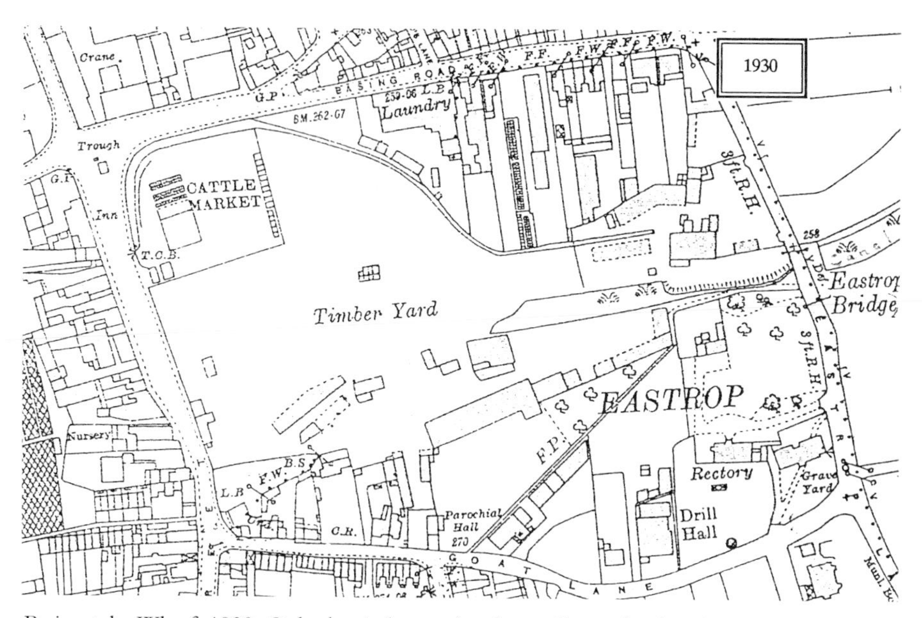

Basingstoke Wharf, 1930. Only the timber yard and a small stretch of reedy canal bed remain. Eastrop Bridge was demolished in 1927 and a culvert substituted beneath the embankment. In 1936 a bus station replaced the timber yard and cattle market

apparently been no other person or body liable to repair them. He thought, therefore, that the attention of the Surrey and Southampton County Councils should be drawn to the Appeal Court's decision, and to the views of the Woking UDC so that an appeal to the House of Lords should be contemplated. So far as the owners of the canal were concerned, the decision of the court was that if the canal was to be carried on as a canal and tolls to be charged, parliamentary powers must be sought. And there the position has been allowed to rest to the present day.

CHAPTER FOURTEEN

AUTUMN DAYS (1913–66)

Of Basingstoke in Hampshire
The claims to fame are small:
A derelict canal
And a cream and green Town Hall.[173]

Last attempt to reach Basingstoke (1913) – a much publicized voyage – compensation for loss of water – trade during the First World War – C.S. Forester – commercial traffic ceases to Fleet (1918), Crookham (1920) and Aldershot (1921) – A.J. Harmsworth takes charge – his early life and subsequent success – barge building at Ash Vale – collapse of Greywell Tunnel (1932) – cessation of coal (1936) and timber (1949) traffic to Woking – publicity aroused by third auction (1949) – a Times leader – the New Basingstoke Canal Company – IWA rally (1962) – death of Mark Hicks (1966).

The *Victoria County History* referred to the canal in 1912 as 'now almost entirely derelict'. A similar description could have been repeated more than half a century later. There was little change. Repairs barely kept pace with decay and longer sections of the waterway began to be abandoned. It was Basingstoke itself which first ceased to use the canal when, after the winding-up of the Hampshire Brick & Tile Company in 1901,[174] coal ceased to be barged from Basingstoke to Up Nately. A letter dated 4 November 1913 from Herbert Hall, wharfinger at Odiham to Alec J. Harmsworth states 'your monkey boat laden with sand from Mytchett arrived in Basingstoke on 18 February 1910'. That was the date of the last barge to reach Basingstoke. A similar voyage was attempted in the autumn of 1913,[175] which attracted considerable publicity in the national press as well as on Gaumont-Graphic (a forerunner of Gaumont-British) News.[176]

The Act of 1778 stated that should the canal be disused for the space of five years, the land should be reconveyed to the previous owners or their successors in title. The Railway & Canal Traffic Act of 1888 also laid down that where, on the application of a local authority, or of three or more owners of land adjoining or near a canal, it appeared to the Board of Trade that the canal, or part of it, had, for at least three years, been disused or unfit for navigation, the Board had power to authorize its abandonment.

In 1913 A.J. Harmsworth made the last attempt to work a barge through to Basingstoke. George Harmsworth is seen at the helm of the narrow boat in Ash Lock with a cargo of 5 tons of sand. The voyage attracted much publicity but the water-level was too low and after three months the boat had only reached Basing

Apparently (as no barge had tied up at Basingstoke Wharf since 1910) it was this threat, and the forthcoming hearing of the company's appeal in the Supreme Court, which promoted Carter to persuade bargemaster Harmsworth to attempt to work a boat through to Basingstoke in the autumn of 1913.

The barge chosen was aptly named the *Basingstoke* and, loaded with nothing more exotic than 5 tons of sand, it left Ash Vale on 16 November. Progress was slow as horse and men towed the craft through the reeds and rushes of the canal's upper reaches but, except for the need to open a stubborn swing-bridge, uneventful, until Slade's Bridge at Up Nately was reached; it was there that lack of water forced the barge to heave to while a dam was built to allow the long pound to Basingstoke to be filled. In the meantime, Carter wrote to Harmsworth on 28 November, after the lawsuit had been won, stating: 'we have our costs for both actions, no liabilities and an assured title in fee simple. So the voyage of the *Basingstoke* loses its significance.' In spite of this news an attempt was made on 8 December to force a passage, but *Daily Express* man Ivor Heald reported the next day that they had been overtaken by bitter ill-luck.

> Little did I think when I sailed away from Mapledurwell last evening, singing shanteys and waving handkerchiefs, that in a few short hours our ship would be coming back this way again, stern foremost. Alas! the canal had another bad puncture during the night, and at dawn the look-out discovered that we were running directly on a mile stretch of dry land. One of the narrowest escapes we have had. The captain, with admirable presence of mind, immediately ordered us to reverse the horse, and we ran back about half a mile for safety. We are now hove to under a tunnel* and, after consulting the chart and repeatedly pacing the distance to the King's Head, I find we have lost about a quarter of a mile as compared with our position ten days ago. It is a cosy sort of tunnel with tophole echoes and very convenient to strike matches on, but somehow I don't like the idea of eating my Christmas plum duff underground. Far better to die a brave death on the open canal than to vegetate in a tunnel and may be get covered with those horrid stalacite things. It was not for this kind of sailoring that I bought a telescope and had anchors tattooed on my arms. . . . As a matter of fact I have made up my mind to desert the ship . . . tonight I shall swim ashore, stain my face with walnut juice, and make my way across country to the nearest British consul.

The proximity of the canal to the Guards' Depot at Pirbright and the army barracks and camp at Aldershot naturally assured its importance in military exercises, where it assumed sometimes the guise of the Rhine or the Siegfried Line or simply a No Man's Land which demanded ingenuity to be crossed. Many a young recruit was grateful for its existence when map reading. It also served for recreation and for debunking colleagues who were pompous or had simply made themselves unpopular. J.R. Colville, Sir Winston Churchill's private secretary, recalls in his biography of Field Marshall Lord Gort VC that when Gort was an ensign in the Grenadier Guards (1905–7) he was far from popular during his early years in the regiment. 'Honest, good-humoured and imperturbable through he might be, there were occasions when his brother officers thought his devotion to duty went altogether too far, and on one

* Little Tunnel.

occasion they threw him into the Basingstoke Canal at Pirbright for taking life too seriously.'[177]

In August 1909 the Frimley & Farnborough District Water Company had obtained a further Act to authorize a well (already partially sunk) and pumping station by the River Whitewater near Deptford Bridge at Odiham, to which the proprietors of the canal made no objection – presumably because Carter was too preoccupied with Bottomley's activities and the Eyre affair pending in the High Court. However, Carter considered that the lack of water in the upper reaches of the canal was mainly due to this well, and claims for compensation were passed which resulted in the Frimley & Farnborough Act of 1915, which provided for the canal to be supplied with water or compensation to be paid if it should be proved that the pumping of the Odiham well had caused any diminution of the supply of water from the springs rising in the Basingstoke Canal within a radius of 3 miles. It also referred to compensation being payable for injury to the canal 'prior to the company having made good such diminution'. A further Act obtained in 1927 also contained a section for the protection of the canal. A study of the nine sub-sections, however, shows that, while in theory the company was protected against loss of water, the difficulties and cost of proving such loss made them virtually ineffectual, although the water company did agree that if weirs and recording appliances were provided by the canal company, it would pay half the cost up to £300 of the work completed within six months from the passing of the Act, excluding delays due to strike, frost, etc.[178]

In April 1914 a new company – The Basingstoke Canal Syndicate Ltd – was formed with a capital of £15,000 and two directors, Harold Irvine and Surman Sibthorp, who apparently held only four shares between them. Six months later £20,000 in debentures were created to pay off the balance of the purchase price and these were redeemable at the option of the new company at 15*s* in the pound before 1 July 1915, which sum increased by 6*d* in the pound for each succeeding period of six months until 1921. William Carter remained the mortgagee.

This latest effort to reopen the canal met with success and within a few weeks of the outbreak of the First World War, on 4 September 1914, a notice appeared in *The Times* under the heading 'Basingstoke Canal Re-opened' which read: 'It should interest all those concerned in the industrial and commercial development of the districts served by the Basingstoke Canal to learn that an ample supply of water has been secured to the canal by the partial diversion of the Whitewater Stream at Odiham, and that the canal is now being used for the conveyance of traffic.' In fact, work on rendering the canal navigable had been in progress since 1912. Two new barges capable of carrying 50 to 60 tons were registered in May 1914 at the offices of the Port of London Authority, and during the summer the bed above Fleet was repuddled with thick clay.[179] It was also in May that William Mills, an out-of-work bricklayer's labourer, drowned himself in the canal at Woking.[180]

During the First World War the waterway was used to convey government stores and munitions from Woolwich to Aldershot as well as such miscellaneous items as beds, clinkers, flour, hop-strings, oats and oil cake. The main down traffic was timber from Fleet and Frimley and horse manure from the camp at Eelmore to the wharves below Brookwood. Control was vested in the Inland Waterways & Docks Department of the War Office and was managed by the Royal Engineers, under Lieutenant Wilder's command, at the Stanhope Lines. At one time twenty-five boats were working and German prisoners-of-war were employed on unloading and maintenance work. The tonnage carried to and from the Wey varied between 11,600 in 1915 and 18,000 tons in 1918. A perusal of the ledgers for 1919 shows that the upward cargoes of the government narrow boats consisted entirely of oats, and that while *Eliza & Anne*, *Elsie & Violet* and *Millie & Minnie* were removing load after load of army boots and rubber tyres, and *Mabel & Enid* carried nothing more lovable than 'old tins'; *Diana* appears to have concentrated on empty drums and scrap metal.

All sorts of people were attracted to the waterway. Canoeist and author R.C. Anderson had spent an interesting ten days in 1906 in a Canadian canoe

German prisoners-of-war unloading timber from *Dauntless* at Frimley Wharf, 1916

on the Wey, Basingstoke, Itchen & Test Navigations, a cruise which he termed 'by no means an overwhelming success' for reasons not narrated, in his *Canoeing and Camping Adventures* (1910). But it was perhaps the solitude and beauty of its reaches which lingered most in memory. During the last days of the First World War, when aged eighteen and sick at heart, author-to-be, C.S. Forester, went off alone camping in a wild wood of birch and fir on the banks of the canal at Brookwood Cemetery. His sojourn there had a profound affect on his state of mind. Fourteen years later he ruminated in his autobiography *Long Before Forty* that

> Perhaps the trees are cut down now, and there are graves and monuments there, but I shall always remember the place as it was then, with the tangle of silver birches and the waterlilies blooming on the canal, where hardly one barge a week came to disturb the solitude. For four weeks I remained there, surrounded by all the camps and barracks of the Aldershot–Woking district, but (thanks to my choice of site) hardly seeing a soul save for the children who sometimes ventured into the wood and who called me the 'Ole Man in the Tent'. At intervals I emerged to buy food and to draw water, but all the rest of the time I was blissfully alone, and climbing back to normality, and realising (what had never occurred to me before) that it was the most foolish thing in the world to worry, and that worry can be controlled if only one sets one's mind on it. I do not think that even I was so priggish as to debate my spiritual troubles with myself in the way the foregoing lines might be taken to indicate; it all worked itself out naturally – fresh air and the cessation of reading two books a day might have had something to do with it too – and I was not conscious of the enormous good that holiday did me until some months later. Four weeks of fresh air, sound sleep, and absence of all contact with the human world made almost a natural young man of me, as was only to be expected, and it was only two months later that the war came to an end and released me from the worst of my troubles and worries.
>
> What those four weeks really did was to give me leisure to sort my thoughts out and distinguish between instinct and reason. The usual interplay of argument with friends does that for most people; thanks to the war I had no friends I could argue with. An hour or two's conversation with and confession to my father might have done as much for me – but my father was lost to sight in the peninsula of Sinai quite as effectively as the Israelites had been, and I had not seen him since I was fourteen. The chasm between fourteen and eighteen is hard to bridge, particularly by a secretive boy always far too preoccupied to write letters.[181]

Another admirer of the canal was author Cyril Connolly whose father lived at the Lock House at Deepcut. He recounted in 1924 how he loved the area.

> I find it incredibly quiet and lovely and very understanding. I am rather gone on it and prowl round the empty rooms or turn on one light or all the lights when my father has gone to bed and go out to the end of the garden to see the golden shafts shining through latticed windows and stretching over the grass to where I stand by the dark yew hedge listening to the soft splashing of our stream, my feet in the dead leaves. I do not know if it is really lovely or if I love it because I have been a child here or because I must love something or because one day it will be mine. It is not a 'place' but rather a large small house, mostly eighteenth century but skilfully added on to and with a lovely garden now rather run to seed, abiding in a hollow by still water and encircled by protecting trees. My father does not like it because it is damp and because it is out of the way, but my mother loves it more than me. It is full of nice creatures, birds that my mother knows all about, squirrels and rabbits and even foxes and very kindly trees. Not as admirable as yours of course, nor have we a vine. We have a stream . . . Water is very thrilling don't you think? . . .
>
> The country is lovely and wild and deserted though I know it well, it is full of variety, sometimes I go for my run at night if my temper has broken up, at night it is very thrilling. We might go out sometimes when you come. If I am tired I go quietly along the canal which is wonderful in the evening.[182]

The 1918 edition of *Bradshaw's Canals and Navigable Rivers* stated that 'the work of reconstruction is in an advanced stage. All the lock gates have been renewed, and the canal has been dredged and is open for traffic to a point six miles beyond Aldershot and will shortly be opened throughout to Basingstoke.'[183] But this was not to be. Captain S.R. Gardner was listed as Manager and Engineer and Herbert Hall as being in charge of the wharf at Odiham.

In 1919 the syndicate went into liquidation and by the end of the following year regular commercial traffic above Woking had virtually ceased. The last load of timber was brought down from Fleet in December 1918, from Aldershot in January 1919 and from Crookham in March 1920. The flour traffic up to Aldershot ceased in October 1919 and that of cut timber to the Farnham Road Wharf five months later. Ash Wharf fell into disuse in 1920 and North Camp in 1921. Receipts for these years are not known but in 1921 the War Office paid the receiver £860.

The passage of twenty-two barge-loads of aeroplane parts during the summer of 1921 not only signified the end of the commercial traffic from Aldershot but also heralded the removal of army flying exercises from Laffan's Plain and recalled the fact that the town had been for many years England's cradle of military aviation. As far back as 1863 a balloon ascent had been made from Queen's Parade to demonstrate their value for military reconnaissance. From the 1890s a school of ballooning flourished. In 1907 the first military airship took off from Aldershot and flew over the canal on its flight to London. The following year bargees must have been surprised to see the first of Samuel Cody's* aeroplane flights from Laffan's Plain, which bordered the north bank of the canal. A flying school was established and oil drums, petrol tanks, aluminium and aeroplane parts became regular cargoes until the opening of Farnborough Aerodrome. Cody even used the canal for testing the floats of a seaplane, built for the *Daily Mail* Coastal Circuit of Britain Contest 1913.

During the early 1920s the receiver obtained £3,780 from the sale of land. The canal changed hands again in 1923 when William Carter sold it to Alexander John Harmsworth, whose family had been connected with the canal since the 1840s and upon which his father and grandfather had spent their working lives.

Harmsworth had an interesting life. He was born the eldest son of a canal carpenter. As a boy he loved boating and spent many hours rowing on the canal. On leaving school he had various jobs as a carter, carpenter and bargeman. Later he passed the oral examination to get his work ticket as a registered lighterman on the Thames Navigation and later still he became a Freeman of the River Thames.

In 1890 Harmsworth had married and lived on a houseboat moored at Ash Vale. His early life was not without hardship. His grandson, Tony Harmsworth, has recounted how his grandfather earned a few extra shillings by sweeping the snow off the ice for skaters during the very cold winter of 1890–1. By dint of hard work, skill and economy, Harmsworth was able to build first one, and then a series of pleasure boats which he let at Ash Vale; so began the popular boat-hiring business which eventually totalled nearly four

* Cody (1862–1913) was a Texan who had come to England in 1896 and acquired British nationality. He was an extremely daring and original character who was in turn cowboy, actor, balloonist and aviator. In 1903 he succeeded in crossing the English Channel in a canvas canoe towed by a huge kite. Three years later he became Chief Kite Instructor at the Aldershot Balloon School. On 16 October 1908 he became the first man in Britain to achieve powered flight by flying the plane he had designed a distance of 463 yd over Farnborough Common. In 1913 he used the canal and Eelmore Flash for balancing tests of the seaplane he had designed for the *Daily Mail* Coastal Circuit of Britain Contest. The area of water was insufficiently large for attempting flight. On 7 August 1913 his career came to a tragic end when he crashed on Laffan's Plain near Aldershot.

Testing seaplane floats on the canal at Aldershot. In 1913 the *Daily Mail* offered a prize of £5,000 to the first pilot to fly around Great Britain in 72 hours without alighting on land. Cody had special floats built by Alec Harmsworth at the Ash Vale boatyard to be affixed to his normal aeroplane. The large central float was loaded with passengers and towed by motor-boat between Great Bottom Flash and Mytchett Lake. The air frame was then carted to Eelmore Flash where the pontoons and float were fitted and the aeroplane's buoyancy and balance proved satisfactory

hundred craft. His houseboat was, in due course, sold to the canal company as a maintenance boat.

Harmsworth's role as a carrier began in 1901 when he bought the *Mabel*, the *Nene* and the *Harriet* from the ill-fated Hampshire Brick & Tile Company at Nately. He also purchased and carried the receiver's stocks of bricks from the bankrupt Nately brick-fields and sold them in Ash and Frimley. Moulding sand, dug at Curzon Bridge, was a similar type of cargo taken to Guildford by water for the Guildford Iron Works at Millmead.

In 1904 the narrow boat *Basingstoke* and the barge *Aldershot* were acquired from the receiver of the Woking, Aldershot & Basingstoke Canal Company and the Ash Vale boat-repairing yard began to come into operation opposite the boathouse. Narrow boats were also renovated in Great Bottom Flash and the remains of the slip and strapping posts can still be seen. By 1905 Harmsworth was in a position to order his first new barge after selling *Mabel* to the Thames Conservancy. The wide boat *Dauntless* was built by Costains of Berkhampstead, was launched on 6 November 1905 and immediately

Derelict reminders of the canal's commercial days: the hulks of narrow boats *Basingstoke, Greywell* and *Mapledurwell* lying in Great Bottom Flash, Ash Vale in 1954, together with an old punt, *The Mudlark*, used for dredging and ice-breaking. At the left can be seen the float of a former seaplane used during the Second World War

proceeded to Paddington to load a cargo of ashes for Woking. *Redjacket* and *Bluejacket* were also built at Berkhampstead in 1909, but *Mapledurwell* and *Greywell* were built in 1912 by Fellows, Martin & Clayton of Uxbridge.

The beginning of the First World War in 1914 created great activity around Aldershot and brought much additional traffic to the waterway. Harmsworth increased his participation in the canal's thriving trade by adding four narrow boats and *Glendower* to his fleet. Two wide boats were procured from Paddington Borough Council in 1915 and 1916 he bought *Reliance* from William Stevens, the owner of the Wey Navigation.

Harmsworth did not enter the barge-building business until 1918 when *Rosaline* was constructed by the Ash Vale boathouse. When two years later he built the tug *Shamrock*, the hull was taken down to Staines for fitting out; to get the head down beneath the bridges, tons of iron billets had to be loaded aboard.

The barges constructed at Ash were to Harmsworth's own design. All were built of English oak with 2½ in. thick Columbian pine bottoms. Most had transom sterns and would load 75 to 80 tons on the Thames and 50 tons to Woking. In 1932 it cost about £900 to build a barge, four to six men being needed for some three and a half months to do the work. The Harmsworth boats were painted with Venetian red gunwhales, coamings, decks and cabin tops. The washboards, bitheads, cabin sides, etc. were mid- Brunswick green. The transom was vermilion, with gold leaf lettering picked out in Prussian blue, with the same colouring for the lettering on the washboards. The barges and some of the narrow boats had a Cambridge blue band round the head from below the outwhale to above the top rubbing band.[184]

After 1920 only pleasure boats and craft, proceeding to Ash Vale for repair, passed beyond Woking although De Salis stated in 1928 that the canal was open to Brookwood and would shortly be reopened throughout.

In 1930 an observer wrote

> long stretches of this waterway are among the quietest and least-frequented places in the country. Rich vegetation abounds on its banks. In places the trees meet above it, forming in summer long tunnels in which the light is tinged a greenish shade, giving to him who may pass through them, in a boat hardly propelled through the weed-thick water, a ghostly and baleful air. Here and there a broken bridge crosses the water, here and there an empty boat-house stands mournfully on the bank. Those who know the canal best are, perhaps, the young men who, in the years immediately following 1914 found it ever and again to be the line of demarcation between the forces of imaginary but contending states.[185]

Greywell Tunnel collapsed in 1932[186] and this, of course, greatly reduced the likelihood of a further revival of traffic to Basingstoke.[187] The consequence of the fall did, however, provide the owners with the opportunity of selling parts of the canal above Greywell, the wharf at Basingstoke being sold in 1936 and subsequently transformed into a bus station. In 1937 they formed the Weybridge, Woking & Aldershot Canal Company, which was voluntarily dissolved in 1950. Although only 9,500 tons were carried in 1921, the Harmsworth family efforts were so successful that the tonnage rose to 21,000 in 1927 and 25,200 in 1935. This was the height of its last revival.

The amount of coal burnt by Woking Gas Company averaged about 14,000 tons a year, and when the company ceased to manufacture its own gas in the 1930s, a substantial proportion of the canal's remaining trade disappeared; by 1937 the traffic carried from the Wey Navigation had fallen to 13,000 tons and by 1940 had dwindled to 2,250. Although it rose to over 4,000 during the next two years, the return of peace brought the virtual

The canal was originally planned to avoid the hill at Greywell by making a long detour around Tylney Hall, but so strong was Earl Tylney's opposition that it was decided to shorten the canal by 7 miles by building the tunnel through Greywell Hill. This view shows the approach to the western entrance in 1920

demise of commercial traffic on the canal. The coal traffic had ceased in 1936, and although it was not until 27 June 1949 that *Gwendoline* carried her last load of 20 standards of timber from the Surrey Docks to Spanton's Yard at Woking, regular deliveries had ended in 1947. Frimley Dock, the main barge-repair depot, had ceased to be used about 1912 and in 1939 was filled in. *Ariel* was the last barge to be built at Ash Vale and was launched in 1935. In 1947 *Perseverance* became the last barge to be winched out of the canal and repaired on the bank. The same year saw the passing of the Transport Act which transferred the majority of waterways in Great Britain from private ownership to the control of the Docks & Inland Waterways Executive of the British Transport Commission. Both the Wey Navigation and the Basingstoke Canal were, however, excluded. On 27 December A.J. Harmsworth died and the Harmsworth family decided to sell the canal.

The news that the canal was to be put up for auction again brought the waterway into the limelight and indeed it was bathed in more publicity than

The western portal of Greywell Tunnel collapsed and landslips have reduced the size of the opening, as this photograph taken in 1967 reveals

at any time in its chequered history. On 25 September 1948 *The Times* wrote a leader 'On Not Fancying Canals', pointing out that there were certain things like small islands, medieval castles and baby giant pandas which, when advertised for sale, made us all become, if only for a moment, would-be purchasers. Conversely, the writer felt that there were some things which, although it was well understood why the owners should wish to dispose of them, it was extraordinarily difficult to visualize anyone wishing to buy, and suggested that the Basingstoke Canal was one of them.

> It is not that this particular canal seems especially undesirable; it is rather that the impact of canals in general on the sales resistance of the average man is so very, very small. As canals go, the Basingstoke Canal, thirty-three miles long and over 150 years old, has a great deal to be said for it. The fact that it was one of the few canals in England to be left out of the Government's nationalisation scheme strongly suggests that there is money to be made out of it. To the Brigade of Guards its sentimental value is considerable; in the Pirbright area it has exerted a profound and, on the

By Order of A. J. Harmsworth, Ltd., and the Weybridge, Woking and Aldershot Canal Co., Ltd.

SURREY AND HAMPSHIRE

Byfleet, Woking, Pirbright, Frimley Green, Ash Vale, Aldershot, Fleet, Crookham, Odiham, Greywell and Old Basing.

PARTICULARS AND CONDITIONS OF SALE OF

The Old Inland Waterway

known as

THE BASINGSTOKE CANAL

running from the Wey Navigation at Byfleet to just above the Hampshire Village of Greywell, a distance of

ABOUT 32 MILES

together with the whole of

THE VALUABLE STANDING TIMBER

and the benefit of certain wayleaves and rents amounting to about

per £345 - 0 - 0 annum

ALSO

VALUABLE RESIDENTIAL AND INDUSTRIAL PROPERTIES

adjoining the Canal and including

SPANTON'S TIMBER WHARF — WOKING

Let on lease to produce **per £80 - 0 - 0 annum.** Tenants paying rates

HOWFIELD COTTAGE — PIRBRIGHT

A detached country cottage let on a weekly tenancy to produce **per £28 - 12 - 0 annum.** Landlord paying outgoings

THE BOAT HOUSE CAFE — ALDERSHOT

Let on lease and producing **per £65 - 0 - 0 annum.** Tenant paying rates

Wharf House — Aldershot

Let on a weekly tenancy to produce **per £39 - 0 - 0 annum.** Landlord paying outgoings

LARGE BUILDERS WAREHOUSE AND STORE

adjoining the above and with the benefit of **Vacant Possession**

Canal Cottage — Crookham

a pleasing old world cottage, let on a service tenancy so that **Vacant Possession** could be obtained

LOCK COTTAGE — ALDERSHOT

a detached bungalow of which **Vacant Possession** could be obtained

THE WHARF HOUSE — ODIHAM

a pleasantly situated detached residence let to produce **per £50 - 0 - 0 annum.** Tenant paying rates

SEVERAL ENCLOSURES OF ALLOTMENT & GARDEN LANDS

BOAT HOUSES at Byfleet, Ash Vale and Fleet

ENCLOSURES OF STANDING TIMBER

All in hand at Fleet, Winchfield, Odiham, etc, which

MESSRS. ALFRED PEARSON & SON

are favoured with instructions to sell by Public Auction, in some 36 Lots, at

The Aldershot Institute, Station Road, Aldershot

on TUESDAY, MARCH 1st, 1949

Commencing at 2 p.m. sharp

(The only Lot for which offers will be considered prior to the Sa e is Lot 1)

Illustrated Particulars and Conditions of Sale may be obtained, Price 4/- per copy, from

The Solicitors:
Messrs. Foster, Wells & Coggins,
Victoria Road, Aldershot.

The Auctioneers:
Clock House Farnborough Tel. 1 (two lines)
and at Fleet, Aldershot and Winchester.

Previous auctions had been in one lot, but in 1949 the waterway was offered in thirty-six. Four lots were withdrawn before the sale and the remainder were sold to various individuals for £16,500 whole, restraining influence on those implacable and easily provoked belligerents, Northland, Southland, Eastland and Westland, and the welcome, if other rather unexpected, sight of its placid, unmistakable waters has for generations done more, probably, to restore the confidence of nervous young officers in their ability to read a map than all the rest of their training put together.

When everything in its favour has been said, however, the Basingstoke Canal still remains for most of us a vision which, although unattainable, is not particularly bright or disturbing. Yachtsmen would have been attracted to a lake, fishermen to a loch, escapists to a lagoon and speleologists to a subterranean watercourse; but it is difficult in the extreme to conjure up an image of the canal-fancier. The type doubtless exists, just as there must be people who love trams, and others to whom the newt is the most endearing of our fauna. And if by any chance none of these wealthy eccentrics comes forward at the forthcoming auction, it would, of course, be possible to internationalise the Basingstoke Canal under a Four-Power Commission comprising those old established riparian states, Northland, Southland, Eastland and Westland; though that, it is to be feared, would be asking for trouble.

The writer's ignorance of the thousands of canal-fanciers throws interesting light on the great growth of interest in waterways in recent decades, and very few of these, I suspect, could be classed as 'wealthy eccentrics'. This leading article created a considerable correspondence and contributions for the canal's preservation dribbled in from India and the distant Falkland Islands.[188] Once again the canal was to become a centre of controversy. The Inland Waterways Association was anxious to prevent another speculator purchasing the canal with the object of selling the canal in lots and stripping the banks of standing timber, which between Fleet and Greywell had been valued at £1,800.[189] As the association's constitution at the time did not permit it to acquire real property, a Purchase Committee was formed composed mainly of members willing to contribute towards the cost. Later it transpired that a majority of this committee believed themselves to be an autonomous organization, not under the control of the IWA, and as a consequence sold it to a third party, the New Basingstoke Canal Company.[190]

The 32 miles of canal bed from Byfleet to Greywell with rents for wayleaves and other rights amounting to £345 p.a., together with the lock-cottages at Frimley, Aldershot and Crookham were sold for £8,000. The remaining thirty-one lots (four were withdrawn before the sale) were sold to various individuals and fetched £8,500. The wharf houses at both Aldershot and Odiham each fetched £900 and Spanton's timber yard at Woking £1,950. The boathouse at Ash Vale made £400 compared with £320 for that at Aldershot where the Boat House Café fell under the hammer for £720.

The lowest successful bids were £12 for a small parcel of freehold land at Woking, £20 for 326 ft of tow-path frontage at St John's Lye and £20 for a parcel of timbered freehold land at Up Nately, being the bed and tow-path of the canal from Brick Kiln Bridge to beyond Penny Bridge. A short stretch of dry canal bed at Basing was bought for £30.

There has been no commercial traffic on the canal since it was bought in 1949 by the New Basingstoke Canal Company. The barges which carried timber to Woking left the waterway at the time of its sale except for the *Glendower* which was converted into a houseboat. Lock XXII at Frimley was partially blown up by the Queen's troops returning from a night exercise in May 1957; the top gate was destroyed and the force of the escaping water caused the gates of the lock below to collapse. The net result of this inconsiderate act was to drain the pound used by the Guards for swimming (the pool was constructed in 1942) and to initiate a lengthy and inconclusive correspondence with the military authorities.

At Easter 1961 a party of boats attempted to ascend the canal and an outboard cruiser managed to reach Pirbright Lock (No. XV), where lack of water prevented further progress. The narrator of this adventure reported that numerous curious spectators were surprised that the locks were workable and 'certainly it was not like cruising on the Thames, as at all but three locks the bottom gate had to be "sheeted" with tarpaulins, and at two absence of upper balance beams required the use of ropes and tackle, while two pounds, one at Sheerwater the other at St John's, required filling.'[191] In 1962 the Home Counties Branch of the Inland Waterways Association held a successful rally at Woking. About thirty craft including four narrow boats were able to assemble below Monument Bridge as a result of voluntary working parties removing some twelve lorry loads of rubbish from the canal. On Easter Sunday – the main day of the event – the weather was perfect. 'In the afternoon the attendance was simply staggering; at one time it was estimated that some 4,000 people were on the wharf watching the boats and the various displays that the local cadets had arranged. Balloons soared merrily skywards; everything edible and drinkable was sold from the refreshment tent; . . . the whole scene, alive with bunting and boats, was one which will long be remembered by all who participated.'[192]

The death of Mark Hicks at the age of ninety-two in July 1966 marked, according to the compilers of the *Guinness Book of Records*, the end of the longest working career in one job in Britain. He started work with the London & Hampshire Canal Company in 1884 at the age of ten and was working as one of the canal bailiffs until four days before his death. For many years he had lived at the wharf cottage at Crookham and his funeral procession fittingly took place along the canal. Little could he have imagined what the future held in store.

CHAPTER FIFTEEN

THE SPIRIT OF REVIVAL (1966–91)

Formation of the Surrey & Hampshire Canal Society (1966) – the case for restoration – attitude of the owners – the great flood of 1968 – publication of London's Lost Route to Basingstoke and Boats from the Basingstoke's Past – death of bargee Alf Rogers (1971) – his reminiscences – trip – purchase of canal by Hampshire County Council (1973) and Surrey County Council (1976) – volunteer labour – the Society's steam dredger – the decauville railway – the boat John Pinkerton – initiation and success of canal cruises – reopening of Ash Lock (1979) – renovation of Frimley Aqueduct (1981) – canal reopened to Greywell (1991) – grand celebrations.

In the course of a debate on future water transport policy in the House of Commons on 18 November 1966, Sir Eric Errington, the rather eccentric and lovable MP for Aldershot, drew attention to the sad condition of the Basingstoke Canal. His forceful speech deploring the worsening state of this and other relics of Britain's great canal age gave added impetus to the formation of the Surrey & Hampshire Canal Society. Indeed, its first public meeting was held the very same day as Sir Eric was pleading its cause in the Lower House and within months the Society could claim more than a thousand members.

Jim Woolgar was the Society's first secretary and it was he and his wife Flo who were mainly responsible for getting the Society off the ground. A Surrey alderman commented that 'they couldn't preserve a couple of tame cats',[193] but this simply reflected the wide divergence of views between the elderly who saw the canal as an abandoned weed-choked ditch threatening pestilence and even causing death, and the younger generation to whom the canal represented part of Britain's industrial heritage which should be restored to order and re-utilized for pleasure-boating, fishing and other recreational activities.

Indeed, one of the Society's first acts was to publish a 56-page booklet entitled *The Basingstoke Canal – the Case for Restoration*. This presented a strong case for removing a public nuisance by converting the old waterway into a public amenity. It was estimated that the cost of restoration to bring it to a reasonable standard of navigation as far as Greywell would be £26,000 using voluntary labour compared with £94,000 if a paid contractor was employed.

The cost, on the other hand, of filling it in might be, they suggested, ten times as much!

The Society's first single-sheet newsletter appeared in January 1967; six-page issues were soon introduced and by no. 54, illustrations had arrived. The one hundredth number was issued in December 1981, the one hundred and fiftieth issue in March 1990. This six-times-a-year newsletter has been one of the reasons for the Society's success. It reports on what has happened and what it hopes or fears is going to happen; it appeals for people, jumble, new members and, above all, for support; it cajoles and coerces its members and their families to assist physically, and many of them do. As a result of the consistently hard work of its editors and helpers it has succeeded in going from strength to strength. Its contributors' regard for historical accuracy is sometimes a little wide of the mark but its editorial policy is as it should be. Dieter Jebens took over as joint editor in 1975 and for twelve years from 1978 was sole editor. His style and approach together with excellent photographs over the years resulted on two occasions in the annual award for the best waterway society publication.

The northern entrance to Basingstoke Wharf from Wote Street, 1904. The iron railings mark the culvert and bank of the River Loddo

The navigation had been in the private ownership of the New Basingstoke Canal Company since 1949 and until 1966 little had been done to improve the navigability of the waterway. Indeed it was rather the reverse, and efforts by the Society to establish cordial relations with the company were coldly received. Offers voluntarily to clear rubbish from the canal bed were ignored. Then came the announcement reported in the *Surrey & Hampshire News* on 13 July 1967 that the Company proposed to replace locks by fixed weirs interspersed with culverted sections so that part of the canal bed could be filled in for property development and car parks. This depressing news provoked a variety of correspondence in the county press from local residents, tow-path ramblers, anglers and boatowners, most of whom strongly disagreed with the company's intentions. Then at a public meeting organized by the Society in Aldershot on 13 November 1969 the chairman, Sidney Cooke, accompanied by Harry Swales, the company's solicitor, made his first public appearance. Both men remained unidentified until question time when Mr Swales introduced himself and his client. Mr Cooke, addressing an expectant audience, clearly misjudged the mood of the meeting when he warned that if the canal was restored there would be boats rushing up and down it. 'Is this what you want?' he asked the packed hall. There was a spontaneous response from the audience: 'yes!', they shouted, rather it seems to Mr Cooke's surprise. Although Mr Cooke did say he would like to see the canal restored, his precise meaning was none too clear since he ventured the view that Ash Embankment could not be safely repaired, which rather indicated that complete restoration was to say the least, unlikely. The Society's newsletter editor commented

> One hopes that Mr Cooke left our meeting in Aldershot a wiser man. He saw a full house representative of a good cross section of public opinion, heard that opinion, and may have realised that the Society was more representative of the general public's thinking with regard to the canal than he had formerly thought.[194]

Certainly the great floods of September 1968 had made the Society's case for restoration even more difficult. During that Sunday night when torrential rain poured down on the Hampshire and West Sussex borders a partially blocked culvert had caused the great embankment at Ash to burst, causing considerable damage to nearby Farnborough airfield. A wooden dam was hastily erected but the canal company had no intention of repairing the breach and, by emphasizing the latent danger in the existing situation, obtained in due course planning consent from Hampshire County Council for the construction of a concrete dam built into the wing walls of Ash Lock.

Tony Harmsworth and Cranley Onslow, MP for Woking, watch the author driving a golden nail into a new lock gate built by volunteer members of the Surrey & Hampshire Canal Society. The ceremony took place at the former barge-building yard at Ash Vale in 1969. The gate – one of a pair of lower gates – was later fitted in Ash Lock with a pair of upper gates also built by volunteers

The early years of the Society's work were indeed fraught with frustration. Because the policy of the owners was to close the navigation, the Society's first task was to fight a determined campaign for public ownership. For over four years the battle flourished and there were times when even the most optimistic members doubted whether the canal could really be reopened as far as Greywell. Nevertheless the Society's membership grew apace; public meetings were regularly held in towns and villages between Woking and Basingstoke. It was rare for any jumble sale or village fête in the vicinity of the canal not to be attended with the Society's stall. 'Save the Basingstoke Canal' car stickers were to be seen far and wide (even on the back of a jeep in Sydney, New South Wales).

What the Society didn't have, and indeed never obtained, was permission actually to carry out restoration work on the canal. The canal company's purpose in owning the canal was never really clear but what was certain was that they did not want the waterway restored. On more than one occasion, police were called to disperse a few zealous individuals who thought a little tow-path clearance might be appreciated. Unable to work on the Basingstoke, members had to satisfy their restoration fervour by working on other canal projects or helping to develop the huge effort which went into the campaign.

Endless meetings were held to involve the public and a monster petition with 15,000 names clamoured for restoration. Any publicity was good publicity – even it seemed the *Daily Mail* feature extolling the delights of living on the canal in a houseboat (at St Johns) which unfortunately sank prior to the appearance of the article. In 1968 the publication of *London's Lost Route to Basingstoke*, in both England and America, provided the Society with a detailed history of the waterway. A few months later appeared the booklet *Boats from the Basingstoke's Past* by Tony Harmsworth, the grandson of the canal's former owner, A.J. Harmsworth, which described in detail some the narrow boats and barges which had once worked on the canal. A particularly interesting feature to tow-path explorers was a sketch showing the location of no less than sixteen boats which had been abandoned over the years between Woodham and the Brickworks Arm at Up Nately.

One of the last surviving bargees to work on the Basingstoke Canal before the First World War, died in 1971. June Sparey was fortunate to record an interview with eighty-year-old Alf Rogers a few weeks before his death.[195] He recalled that his first trip on the canal had been about the turn of the century when he was nine or ten years old. About 1907 his step-brother Dick Cobbe (who died in March 1955 aged seventy-three), decided to leave the Thames on which he worked and move to Basingstoke. 'If you like,' he said to Alf 'you can come with me as a cabin boy aboard one of the tugs and we'll see how you take to it'; 'I took to it all right,' Alf remembered.

The first trip on 'the cut' that he recalled was taking a boat ('no. 12 it was') down to Shadwell to pick up a load of coal for the Woking Gas Company – about 40 tons. In those days, of course, the barges were unpowered. 'You pushed, shoved, got a sail up and got there. Horses towed you to Weybridge, and after that you had to use your wits to get to Kingston. Dick and I were the only ones who ever braved flood water on the Thames to get to Kingston. Once you got there, a tug picked you up. Coming back you had a tug all the way to Weybridge. Horses were no good for pulling you up the Thames loaded.'

Alf first skippered the *Dauntless*. Then came the *Redjacket* (1909) and the *Bluejacket* (1911). 'That was the one that always went down to the Tilbury part of the Thames, and it's a nasty bit of work when you take a canal barge down there.' He also captained the *Aldershot*. Alf described A.J. Harmsworth, the canal's former owner, as a man 'who would always take his coat off and get down to it if a job needed doing'. In those days, the barges carried a general cargo – mostly coal and timber – and because the canal was in quite good order climbing the fourteen locks at Frimley was 'no trouble'. It used to take about one-and-a-quarter hours to get through the six locks at Woodham fully loaded. The run from Woking to Tilbury was done in forty-eight hours 'regardless', and that included getting back to Woking with a load. Problems of water shortage only occurred in summer.

The barges operated with a skipper and a mate. Woolwich was the main place to go for coal. Alf recalled that at Shadwell, just outside St Catherine's Dock, a 'dear little monkey boat nearly sank. It took 45 ton of coal which made its deck level with the water'. He also remembered how in 1913 A.J. Harmsworth, Dick Cobbe and himself tried to get a barge through to Basingstoke; 'We had a block and tackle ourselves over the sand heaps. We got as far as Greywell* but couldn't get any further.'

When Alf married, his wife joined him on the barges. The Harmsworth boats had two side bunks in the forward cabin and stools. 'Captains had the aft cabin because the cooking utensils were there. The mate had to go for'ard whether he liked it or not.' Cobbe and his family almost lived on the Thames and the Basingstoke. Space was short; the aft cabin bunks folded up into the wall, and there was always the uncertainty of your whereabouts – 'you didn't know whether you were coming back on the next tide or next week'.

The pay was about 35*s* a week. 'That was good money. But you had to be a good man to get it.' In 1915 Alf's association with the Basingstoke ended when he joined the Royal Flying Corps.

It took four years of incredibly hard work before Hampshire and Surrey County Councils agreed to open talks with the canal owners to discuss terms for an early purchase. It was July 1970, but it would be a further six years before the purchase was completed. The following year, the Earl of Onslow became the Society's first president and even with his substantial help, the negotiations dragged on throughout the year. A major problem was valuation, the owners indicated that the canal was worth £100,000 while many people considered a tenth of that sum a more realistic figure. After a further year of fruitless negotiation, the Society urged both county councils to end the impasse and seek a compulsory purchase order. In 1973 – the year when the Canal Society's membership reached 2,000 – both Surrey and Hampshire County Councils decided to do so and a public enquiry was held. No one objected to the Hampshire County Council's order but 140 objections were

* The *Basingstoke* probably reached Basing village, see p. xx.

lodged against that of Surrey County Council. An incredible amount of nonsense was voiced by objectors ('Boats must not be permitted to use the canal and spoil it', etc.) but to the enormous satisfaction of almost everyone, the orders were confirmed and in November, Hampshire County Council acquired 16 miles of waterway.

In November 1973 the first official voluntary working party began clearing the tow-path at Barley Mow Bridge. The equipment obtained by the Society could at least be put to effective use. The 70 ton steam-powered dredger, bought from the Kennet & Avon Canal, was restored and started work at Colt Hill the following year. The fact that it took ten years for the voluntary dredger team to clear 7 miles of waterway between North Warnborough and Dogmersfield gives an indication of the tremendous amount of mud and debris which had to be removed in skips hauled along a decauville railway laid along the tow-path. As the dredger progressed westwards the narrow gauge line was extended, sidings added and a diesel locomotive introduced to relieve aching backs. The track extended 300 yd and an eight-skip train could carry 10 tons to the dump site. In this manner 70 yd of canal were dredged each weekend. Later, when a tug with a dragline crane and ballast barge were introduced, the track was transferred to Deepcut to carry construction materials between locks XXII and XXVIII. Later still, in 1980, the railway was moved to Ash Embankment where it was used to transport 14,000 tons of clay needed to repuddle the canal bed.

In 1975 European Architectural Heritage Year inspired Hampshire County Council to restore the eastern portal of Greywell Tunnel. Members of the Society played their part by responding to the appeal to buy a brick for 20p. Nearly 5,000 bricks were 'sold' and seven bridge parapets neatly capped as a result. The major development in 1976 was the completion of the purchase of the eastern half of the canal by Surrey County Council for £40,000. Thirty-two miles of the canal were now in public ownership. It had taken nine years.

A traffic survey was undertaken by the Society to determine the amount of use of the North Warnborough Lifting Bridge. After much research it was found that average daily traffic amounted to 4 cars, 2 bicycles and 9 pedestrians. An appeal to raise funds for a cruise boat excited many generous people and the following year a hull was purchased and fitting out began.

The first lock was fully restored in 1977. The 'Deepcut Dig' brought 600 volunteer navvies to the flight. They got through an amazing amount of work and 1,400 pints of beer. Work on the canal was increased significantly when over £35,000 was provided under a six month Job Creation Programme to assist with the restoration of the Deepcut flight. This assistance under the Manpower Services Commission job-training scheme continued for seven years until 1984 and together with other grant aided contributions a total of over £750,000 was expended in restoration work.

The passenger boat *John Pinkerton* passing the section of the canal at Dogmersfield which had always suffered from slippage. In 1983 two JCBs were used to provide a 'final solution'

In 1976 the Canal Society decided to raise £10,000 to build a pleasure boat for canal excursions. The 67 ft long *John Pinkerton* was designed to have the appearance of a traditional narrow boat although its actual width is 8 ft 6 in. to provide capacity for up to 56 passengers. Its headroom of 5 ft 3 in. and draught of 1 ft 10½ in. took into account the low headroom of some of the present bridges. In fact it has less than 1 in. to spare when passing under Reading Road Bridge in Fleet. The maiden voyage took place on 20 May 1978 from Colt Hill Wharf to King John's Castle. Initially it was hoped to make an annual profit of £3,000, but the popularity of the cruises has exceeded all expectations. The average profit for the ten years 1978–87 was over £11,000 p.a. and in 1990 totalled £20,500.[196] The *John Pinkerton* has over the years gradually extended its cruising range; it was first able to descend the Deepcut flight in 1985 and reached the Wey Navigation for the first time in May 1991.

During 1981 Frimley Aqueduct was renovated by British Rail and by the

The reopening of Cowshott Manor Bridge, Pirbright, October 1982. Left to right: Frank Jones, Robin Higgs, chairman of the Surrey & Hampshire Canal Society, the author and Councillor Evelyn, High Sheriff of Surrey

end of the year twelve locks had been restored and 24 miles of channel dredged. In 1983 the derelict dry dock at Deepcut, filled in during the 1930s, was restored. In 1984 the first rally of boats on the canal for over twenty years was held at Ash Lock and in May 1986 the Mayor of Woking formally reopened the Brookwood flight of three locks.

An unexpected development occurred in March 1987 when the Nature Conservancy Council* revealed plans to designate 25 miles of the canal as a Site of Special Scientific Interest and to press for the limitation of motorized boating. Three years later the council further proposed to ban motorized boating west of Odiham. The Society was quick to condemn the proposals on

* The National Conservancy Council is the body responsible for advising Government on nature conservation in Great Britain. Its responsibilities include the management of nature reserves and the identification and notification of Sites of Special Scientific Interest.

The Basingstoke Canal
ROYAL REOPENING CELEBRATIONS

THE BASINGSTOKE CANAL REOPENING
JOHN PINKERTON
VALUE ONE SHILLING
10th MAY 1991 FRIMLEY LODGE PARK

Programme of Events
Frimley Lodge Park, Friday 10th May 1991

ADMISSION FREE Public car park £2

The programme cover for the royal reopening celebrations, 10 May 1991 showing the junction of the Wey Navigation and Basingstoke Canal the grounds that the canal was restored to reopen a public navigation and recreational amenity and not to create a nature reserve. Sir Cranley Onslow, MP for Woking and a vice-president of the Society was prompted at Nicholas Ridley's suggestion to write to the chairman of the Nature Conservancy Council drawing his attention to

> the absurd situation that appears to be developing whereby your Council seems set on thwarting all the efforts that have been put in, over the past twenty years, to restore the canal to navigation. As you will be aware the Surrey section of the canal, almost all of which runs through my constituency has been virtually derelict until fairly recently and could scarcely have been classed as a 'freshwater habitat' if it had not been for the work of which the canal society has so laboriously carried out.

In 1988 the first boats were able to navigate from Hampshire to Woking following the restoration of the St John's flight of locks; the footbridge which had until the 1950s linked the canal at Woodham with the Wey Navigation was also reinstated. By July of the following year all the Surrey locks had been rebuilt; only the completion of the dredging programme and a plentiful supply of water were now awaited. After so many years of effort it was frustrating that 1989 and 1990 were relatively dry. Then the rains did arrive and only the date for the official reopening waited to be set.

On 10 May 1991 HRH The Duke of Kent visited Frimley Lodge Park and paid fitting tribute to the achievements of those many people who had over more than two decades revitalized the old canal. He was accompanied by Robin Higgs, chairman of the Canal Society, and they and a host of other dignatories then boarded the *John Pinkerton* for a ceremonial cruise. The grand celebrations, attended by over 1,000 visitors, continued throughout the day and culminated with a fine fireworks display and a procession of gaily illuminated boats. What now lies ahead?

CHAPTER SIXTEEN

FUTURE OUTLOOK

The Basingstoke Canal Authority – Hampshire County Council purchases canal from Greywell Tunnel to Penny Bridge (1990) – The Surrey & Hampshire Canal Society's engineering proposal for restoration – the bat problem – opening of the Basingstoke Canal Centre (1993).

The Basingstoke Canal, restored to Greywell after many years of dedicated effort, has become the responsibility of the Basingstoke Canal Authority which was formed by the Surrey and Hampshire County Councils in 1990. Its main task is to ensure that this attractive amenity will continue to provide public recreational facilities. The authority coordinates the interests of both county councils, the riparian district councils and representatives of the special interest groups such as the Canal Society, the angling associations, naturalists and ramblers. The main problems facing Paddy Field, its first director, are not unlike those faced by earlier canal managements. They are to ensure an adquate water supply throughout the year, regular maintenance and facilities for boating, walking and car-parking. Equally, he must be aware of any outside activity which might have an adverse effect on the waterways such as building development and vandalism, and act accordingly.

The reopening of the canal in 1991 was followed by a severe water shortage which soon resulted in the closing of the Woking section and, later in the year, Ash Lock. Consequently additional water supplies have had to be obtained by building a well at the Rive Ditch at Sheerwater and by erecting a pumping station at Frimley Green towards whose cost the Society contributed £10,000. Indeed the most serious problem facing the Canal Authority and the Society is keeping the restored length of waterway navigable. Not only must the water supply be increased but dredging must be continued and sites found for the disposal of silt.

It has always been the Surrey & Hampshire Canal Society's aim to restore the canal through and beyond Greywell Tunnel.[197] In May 1986 the Society produced a draft document which created sufficient interest to encourage it to put in hand a study to ascertain what engineering work would be required. The two-volume report, *The Basingstoke Canal – Greywell Tunnel to Mapledurwell Restoration*, issued in 1989, set out in great detail the task which lay ahead. The study, the culmination of many months of thorough research

Brick Kiln Bridge, Mapledurwell, 1993. Restoration is now in progress as far as Penny Bridge

by Stan Mellor and his colleagues, identified the possibilities and did not shrink from spelling out the environmental issues which restoration to full navigational standards would probably arouse. In 1990 the Society published a booklet, *Basingstoke Canal: The Promise of the Western End*, which summarized its proposals. In the same year Hampshire County Council purchased for £10,000 from the New Basingstoke Canal Company, the land forming the bed and tow-path of the canal from the western end of Greywell Tunnel to the site of Penny Bridge, Mapledurwell.

The present (1993) limit of navigation is the Whitewater Winding Hole just west of the aqueduct. The 800 yd to the eastern entrance of the tunnel has been dredged and the site of lock XXX, whose rise and fall was seldom more than a foot, tidied up. Careful examination of the tunnel by the Society's special project group has revealed that the spoil from the roof falls stretches over a distance of about 575 ft along the tunnel or about one-sixth of its total length. At the western end the portal has completely collapsed leaving only a hole in the hill some 4 ft high. The cutting beyond the tunnel was known as The Slip as a consequence of the mud slides which first occurred within six weeks of the opening of the canal in 1794 and which completely blocked the channel. Subsequent slippages happened from time to time throughout the nineteenth century. Four hundred yards further west stands Eastrop Bridge, which is still in service for pedestrians and light traffic. The bed beyond the bridge continues to hold water but was until 1993 choked with trees, both upright and fallen. Slade's Bridge, ¼ mile on, is also in regular use and has a metalled surface; one parapet has been replaced by steel railings. The Brickworks Arm built in 1898 was formerly crossed by a timber footbridge which collapsed in the 1960s. Brick Kiln Bridge, which carries the road from Up Nately to Nately Scures, was purchased by Hampshire County Council *c.* 1924 and remains in fair condition. The channel and tow-path to Penny Bridge have recently been cleared of trees and undergrowth. Penny Bridge itself was demolished in the 1920s and the Greywell to Hatch road now crosses the canal bed on a new alignment.

Restoration beyond Penny Bridge would ensure that Little Tunnel, a grade 2 listed structure, was repaired but the building of the six-lane M3 motorway has destroyed all trace of the canal at Hatch, cutting off any immediate hope of reopening the canal past Basing House to Basingstoke.

The main constraint in restoring the tunnel is neither cost nor the engineering difficulty, but bats. Bats are a protected species (under the Wildlife and Countryside Act (1981)) and have for some years become dependent on the tunnel as a winter resort. Only in 1985 was it realized that a significant number of them hibernated there when, on 24 January, over five hundred were counted in the eastern end. As might be imagined, this unexpected discovery has been the subject of much discussion by Society members who have, I believe rightly, challenged its degree of importance having regard to the fact that the tunnel was built for navigation. The position has now been investigated by Dr R.E. Stebbings who in June 1992 reported on behalf of the South Region of English Nature on the potential effects of the bat population if the tunnel was reopened. This ninety-page document[198] concluded not only that the tunnel had more bats than any other canal tunnel in Britain but also that more bats were to be found in the tunnel than anywhere else in the country. Commenting on the Canal Society's proposal to restore the tunnel over a period of time, he felt that overall if the scheme to restore the tunnel was implemented (including the provision of a purpose-made tunnel), some bats would be killed in the course of the work and significant numbers of others would be deprived of a roost. This might affect their survival and a severe decline in the number of bats in the area could be expected. There was, he felt, no evidence to suggest that bats, in large numbers, would be able to continue using the area for hibernation. Indeed, the proposals if implemented would, Stebbings considered, be likely to lead to a substantial decline in bats, even with amelioration tunnels. The cost of constructing a large enough tunnel for bat hibernation would be in the region of £250,000 but such a feature could not be guaranteed to attract substantial numbers of bats. Discussions between the interested parties continue but it is not evident at present what the new owners of the canal will decide to do. In the meantime work is progressing on clearing and maintaining the bed and tow-path from the western end of the tunnel to Penny Bridge.

The need to ease traffic on the Farnborough Road between the M3 at Frimley (junction 4) and the A31 at Farnham led to various proposals for the Blackwater Valley Relief Road which would allow a dual-carriageway to pass either under or over the canal. If it passed over there was the likelihood that six locks would be needed to allow water traffic to pass beneath. As there are already more locks on the canal than many would wish, it was good news when it was decided to provide an aqueduct. However, agreement has not yet been reached on the design, and work scheduled to begin in the autumn of 1993 has been delayed.

Volunteer work started in 1992 to recreate the towpath and the line of the canal to Penny Bridge as part of a proposal called 'The Last Five Mile Project' by the Basingstoke Heritage Society. This scheme seeks to recreate a walk along the canal line, as far as possible, into the centre of Basingstoke, thus creating a long-distance walk to London via the Basingstoke Canal towpath, the River Wey Navigation towpath and the River Thames towpath.

A recent improvement has been the establishment of a canal centre on the banks of the waterway at Mytchett by Surrey County Council. The centre, besides acting as a focal point for canal users, houses the authority's offices. It also provides an information centre including an interesting collection of historic photographs and artefacts as well as a meeting room and bookshop. It was opened on 22 May 1993 by the chairman of Surrey County Council, Sandy Brigstocke.

The banner headlines of the *Basingstoke & North Hampshire Gazette* on 17 September 1993 read 'BID TO REOPEN TOWN'S CANAL. £10 Million Vision To Kill Grey Image And Attract Visitors And Business'. This latest development relates to a proposal by the former Mayor of Basingstoke, Keith Chapman, to give the town centre a new focal point. Subsequently the Basingstoke and Deane Borough Council commissioned a feasability study to assess the engineering requirements and constraints. A cost benefit analysis may demonstrate that a restoration project could be economically viable. It would certainly increase the town's attractions and I, for one, would not be surprised if one day barges once again moor where buses now wait.

APPENDICES

1. Table of Locks and Distances from London Bridge to Basingstoke, 1900

Navigation	*Lock*	*Place*	*Distance* (*miles flgs*)	*Distance from London Bridge* (*miles flgs*)	*Height above datum at Liverpool at head of locks* (*feet*)
Thames		From London Bridge to:			
(30 miles)	1	Teddington	18 5½	18 5½	24
		Kingston Bridge	1 6	20 3½	
		Hampton Court Bridge	2 7	23 2½	
	2	Molesey	0 1½	23 4	30½
	3	Sunbury	2 7½	26 3½	36
		Entrance to Wey Navigation	3 5½	30 1	
Wey	4	Thames	0 2	30 3	46
(3 miles)	5	Weybridge	0 5	31 0	54½
	6	Coxes	0 5	31 5	59½
	7	Newhaw	0 7	32 4	68
		Junction Basingstoke Canal	0 5	33 1	
Basingstoke	8	Woodham I	0 2	33 3	75
(37 miles)	9	Woodham II	0 3	33 6	81
	10	Woodham III (lock-house)	0 4	34 2	87
	11	Woodham IV	0 2	34 4	93
	12	Woodham V (Sheerwater Bridge)	0 1	34 5	100
	13	Woodham VI	0 6	35 3	105½
		Monument Bridge, Woking	0 6	36 1	
		Woking Gas Works	0 1	36 2	
		Spanton's timber yard	0 2	36 4	
		Wheatsheaf Bridge, Woking	0 4	37 0	
		Hangdog Bridge	0 4	37 4	
		Arthur's Bridge	0 2	37 6	
		Tramway from brickworks	0 0½	37 6½	

Navigation	*Lock*	*Place*	*Distance* (miles flgs)	*Distance from London Bridge* (miles flgs)	*Height above datum at Liverpool at head of locks* (feet)
Basingstoke (cont.)	14	Goldsworth VII	0 4½	38 3	112½
	15	Goldsworth VIII	0 0½	38 3½	119½
	16	Goldworth IX (lock-house)	0 1½	38 5	126½
	17	Goldsworth X	0 0½	38 5½	132½
	18	Goldsworth XI	0 0½	38 6	138½
		St John's Bridge, Woking	0 0½	38 6½	
	19	Brookwood Bridge XII	1 5½	40 4	145½
	20	Brookwood XIII	0 0½	40 4½	152
	21	Brookwood XIV	0 1	40 5½	158
		Pirbright Wharf (lock-house)			
	22	Frimley XV	1 0½	41 6	165
	23	Frimley XVI	0 2	42 0	171½
		Cowshot Manor Bridge	0 1½	42 1½	
	24	Frimley XVII	0 1	42 2½	178
	25	Frimley XVIII	0 1½	42 4	185
	26	Frimley XIX	0 1	42 5	191½
	27	Frimley XX			199
	28	Frimley XXI	0 0½	42 5½	206
	29	Frimley XXII	0 0½	42 6	212½
	30	Frimley XXIII	0 0½	42 6½	219½
	31	Frimley XXIV	0 0½	42 7	226
	32	{ Curzon Bridge Frimley XXV	0 1	43 0	233
	33	Frimley XXVI	0 1½	43 1½	239½
	34	Frimley XXVII	0 1	43 2½	247½
	35	Frimley XXVIII (lock-house)	0 2½	43 5	252½
		Frimley Dock	0 0½	43 5½	
		Entrance Deepcut	0 0½	43 6	
		Exit Deepcut	0 4½	44 2½	
		Frimley Aqueduct	0 5	44 7½	
		Frimley Green	0 0½	45 0	
		Mytchett Lake	1 3	46 3	
		Ash Vale Railway Station	0 4	46 7	
		Ash Wharf	1 2	48 1	
		Gas Works	0 7½	49 0½	
	36	Ash XXIX (lock-house)	0 0½	49 1	260½
		Aldershot Bridge & Wharf (for Farnham)	1 6	50 7	

Navigation	*Lock*	*Place*	*Distance* (*miles flgs*)	*Distance from London Bridge* (*miles flgs*)	*Height above datum at Liverpool at head of locks* (*feet*)
Basingstoke (cont.)		Eelmoor Bridge	1 5	52 4	
		Pondtail Bridge, Fleet	1 2	53 6	
		Reading Road Bridge and Wharf, Fleet	1 0	54 6	
		Crookham Wharf	2 0	56 6	
		Barley Mow Bridge (for Winchfield)	3 0	59 6	
		Pillers Bridge	1 0	60 6	
		Broad Oak Bridge	0 3	61 1	
		Odiham Bridge & Colt Hill Wharf	0 5	61 6	
		Swan Bridge, North Warnborough	1 1	62 7	
		{ Odiham Castle			
		{ Whitewater Aqueduct	0 4	63 3	
	37	Greywell XXX	0 3	63 6	261
		Greywell Tunnel (East End)	0 1	63 7	
		Greywell Tunnel (West End)	0 5½	64 4½	
		Entrance Nately Brickworks	0 1½	64 6	
		Brick Kiln Bridge, Up Nately	0 3	65 1	
		Little Tunnel	0 3½	65 4½	
		Mapledurwell Swing-Bridge	0 2½	65 7	
		Hatch Bridge	1 0	66 7	
		Basing Wharf	1 1	68 0	
		Basing House Bridge	0 1	68 1	
		Slaughter Bridge, Basing	0 2½	68 3½	
		Wellocks Mill	0 2½	68 6	
		Eastrop Bridge	1 1	69 7	
		Basingstoke Wharf	0 1	70 0	

Max. size of barges	72 ft 6 in. × 13 ft to Basingstoke, 72 ft 6 in. × 13 ft 10½ in. to Ash Vale
Draught	4 ft 6 in.
Min. headroom	9 ft 6 in.
Length of tunnels	Greywell 1,230 yd (16th longest in Great Britain), Little Tunnel 50 yd

2. THE COMPANY OF PROPRIETORS OF THE BASINGSTOKE CANAL NAVIGATION, 28 MARCH 1788

A

Mr Lancelot Atkinson, Market Street	500

B

Mrs Elizabeth Ballard, Duke Street, St James's	300
The Corporation of Basingstoke	500
Alexander Baxter, Esq., Kensington	300
Alexander Baxter, Esq., Odiham, Hants	600
William Bayley, Esq., Warwick Street, Charing Cross	1,000
Mr John Beard, Market Street, St James's	500
Mrs Elizabeth Bell, Saville Row	100
Mr William Bent, St Martin's Lane	1,000
Mr Daniel Bergman, Charles Street, Grosvenor Square	500
Mr Charles Best, Basingstoke, Hants	300
Dr Robert Bland, St Alban's Street	300
Mr Richard Bloxham, West Dean, Wilts	300
Mr Robert Blunt, Odiham	100
Mr Richard Booth, Worting, Hants	100
Mr William Nevill Brockett, Budge Row	100

C

Mr John Campa, Park Lane	500
John Child, Esq., Mount Row, Lambeth	100
Miss Eleanor Robinson Child, Ditto	200
Miss Science Child, Ditto	100
Thomas Lobb Chute, Esq., Vine, Hants	500
Mr John Covey, Basingstoke	300
Mr Thomas Creswell, St James's Square	500
Mr John Crowson, Charles Street, St James's Square	200

D

Right Hon. Earl of Dartmouth	4,000
Mr John Davies, Basingstoke	300
Mr Benjamin Davies, Reading, Berks	100
Mr William Davies, Henrietta Street	500
Mrs Elizabeth Davies, Bagington, Coventry	200
Philip Dehany, Esq., Queen Anne Street, Cavendish Square	500
Mr Anthony Demezy, Hartley Row, Hants	100
Mr Richard Doyle, No. 75, Berwick Street	200
Revd Dr John Duncan, South Warnborough, Hants	2,800
Revd Dr Duncan's Four Children	400

E

John Edwards, Esq., Worting	1,000
James Esdaile Esq., Duke Street, Westminster	500
Sir James Esdaile, Kt. and Alderman of London	300
John Esdaile, Esq., Bunhill Row	300

F

Revd Mr James Filewood, Stifford, Essex	500
Revd Mr Christopher Fox, Cliddesden, Hants	300
Mr Thomas Fynmore, Aldersgate Street	300

G

Anthony Gander, Esq., Secretary of State's Office	300
Mrs Bridget George, Tyler Street	200
Mrs Eleanor Glover, Albemarle Street	200
Mr David Graham, Basingstoke	100
Mr John Granger, Weybridge	300
The late General Francis Grant	300
Mrs Mary Griffiths, Stanhope Street	200
Mr John Griffiths, Stanhope Street	300

H

Mr Thomas Hack, Basingstoke	100
Thomas Hall, Esq., Preston, Candover, Hants	1,000
Dr James Hardy, Northampton	300
Mr William Hart, Chapel House, Enfield	200
John Harwood, Esq., Dean, Hants	300
Mr Thomas Hasker, Chinham, Hants	500
John Hingestone, Esq., King's Street, Cheapside	200
Christopher Hodges, Esq., Bramdean, Hants	300
Mr William Hodgkinson, Daventry, Northamptonshire	300
William Holden, Esq., Birmingham	2,000
Revd Mr John Rose Holden, Queen Street, Westminster	1,000
Richard Holland, Esq., Weybridge	200
Mr Humphry Hutchins, Earls Court	500

I

Mr William Irish, St Alban's Street	500

J

Walter James James, Esq., Upper Brook Street	2,000
Mr Thomas Jarvis, Charing Cross	1,000
Thomas Jefferys, Esq., Cockspur Street, Charing Cross	2,000
Richard Jeffreys, Esq., Basingstoke	1,000

K

Mr William King, Andover	500
Mr John Kirby, Downhursbourn, Hants	100

L

Mr Elisha Lane, Walton upon Thames	300
Bennet Langton, Esq., Queen Square, Westminster	1,500
Mr George Langton, Queen Square, Westminster	300
Mr Thomas Leech, Upton Gray, Hants	100
Revd Mr George Lefroy, Ash, Hants	500
Samuel Licnigary, Esq., Layton, Essex	600
John Lovett, Esq., Overton	500
Mr John Lyford, Basingstoke	300

M

Robert Mackreth, Esq., Ewhurst, Hants	1,000
Alexander Mair, Esq., Kensington	300
Mr Francis Martelli, Sherborn St John, Hants	100
Mr Robert Mawley, Tottenham Street	200
Henry Maxwell, Esq., Ewshott	500
Mr Thomas May, Basingstoke	300
Mrs Agatha Mayes	100
Mr Joseph Mollotte, Stanhope Street	100
Mr John Monkhouse, Jermyn Street	1,000

N

The late Right Hon. Earl of Northington	2,000
Andrew Newton, Esq., Litchfield	4,000
Mrs Mary Newton, Litchfield	500
Mr Thomas Nicholls, Princes Street, Cornhill	100
Samuel Nicolls, Esq., Hatton Garden	1,000
Revd Mr Henry Norman, Morestead, Hants	100
Mr Percival North, Fleet Street	100
Mr Frederick Nutt	500

P

Right Hon. Earl of Portmore	1,000
Right Hon. Earl of Portsmouth	4,000
Mr Samuel Palmer, Birmingham	500
Benjamin Parker, Esq., Birmingham	1,000
David Parker, Esq., King's Mews	500
Mr Joseph Parker, Whitney, Warwickshire	300
Mr George Penton, Basingstoke	300
Messrs William Pink and William Knight, Basingstoke	300

Joseph Portal, Esq., Freefolk, Hants	2,000
George Powlett, Esq., Amport, Hants	1,000
Mr John Prothero, Charles Street, St James's Square	300

R

Right Hon. Lord Rivers	4,000
Alexander Raby, Esq., Cobham, Surrey	300
Mr Thomas Rendall, Dean Street, Soho	300
Mr Thomas Richardson, Odiham	100
Mr John Ring, Basingstoke	100
Mr William Ring, Basingstoke	100
Mr Thomas Robins, Basingstoke	300
Revd Mr Robinson, Burghfield, Berks	500
Mr Marturen Luis Roche, Stanhope Street	300
Revd Mr William Rose, Daventry	1,000
Mrs Mary Rose, Charles Street, St James's Square	100
Mr Samuel Russell, Birmingham	300
Mr William Russell, Basingstoke	100

S

Sir Henry Panlet St John, Bart., Dogmersfield, Hants	1,000
Revd Mr St John, Odiham, Hants	300
Thomas Salt, Esq., Birmingham	500
Revd Mr Edward Salter, Strathfieldsaye, Hants	100
Mr Thomas Saunders, James Street, Golden Square	300
George Scott, Surveyor, Thavies Inn	300
Revd Dr Thomas Sheppard, Basingstoke	300
Mr James Sheriff, Birmingham	300
Mr Samuel Shipton, Reading, Berks	100
John Slade, Esq., Hammersmith	1,000
John Lewin Smith, Esq., Grosvenor Street	2,000
Mr John Smith, Pall Mall	400
George Stainforth, Esq., Weybridge, Surrey	300
Mr Richard Stainforth, Weybridge, Surrey	300
Simon Stephenson, Esq., New Way, Westminster	300
Mrs Steward, Wilson Green, near Birmingham	300
The late Revd Mr Thomas Stockwell	100
James Wallis Street, Esq., Bucklesbury	500
George Stubbs, Esq., Suffolk Street, Charing Cross	400
Messrs Thomas and George Stubbs	200

T

Mr Samuel Toomer, Basingstoke	300
William Turner, Esq., Birmingham	500
Mrs Sarah Turner, Birmingham	300

V

Mr Peter Vere, Kensington Gore	500

W

Mr Charles Wade, Crane Court, Fleet Street	500
Nathaniel Wakeford, Esq., Pamber, Hants	300
Mr John Walther, Charles Street, St James's Square	200
Mr Charles Ward, Kensington Gore	600
The late Revd Mr Benj. Webb	300
Mr James White, Attorney, Chancery Lane	100
Mr Thomas Williams, Bull Wharf, Queenhithe	300
Henry Wilmot, Esq., Aldershot, Hants	300
Charles Godfrey Wolff, Esq.	300
Mrs Wood, Highfield Place, near Farnham	300
Mr Richard Wright, Charles Street, St James's Square	300
Mr Joseph Wright, Charles Street, St James's Square	100
	£86,000

3. Proprietors of the Basingstoke Canal, 1788–1994

1788–1866	Basingstoke Canal Navigation Co.	(see Appendix 4)
1866–1874	in liquidation	Frederick Whinney, receiver
1874–1878	Surrey & Hants Canal Co.	William St Aubyn, chairman
1878–1880	in liquidation	Mr Laulman, receiver
1880	J.B. Smith	
1880–1882	Surrey & Hampshire Canal Corporation Ltd	Edward Dawson, chairman
1882–1883	in liquidation	F.G. Painter, receiver
1883–1887	London & Hampshire Canal and Water Co. Ltd	Edward Dawson, chairman
1887–1895	in liquidation	F.G. Painter, receiver
1895–1896	Sir Frederick Hunt	
1896–1900	Woking, Aldershot & Basingstoke Canal & Navigation Co.	Sir Frederick Hunt, chairman
1900–1905	in liquidation	P.E. Gauntlett, receiver
1905	William Carter	
1905–1906	Joint-Stock Trust & Finance Corporation	Horatio Bottomley, chairman
1906–1908	in liquidation	D. Eason, receiver
1908–1909	London & South-Western Canal Co.	William Carter, mortgagee
1909–1914	in liquidation	H. Brougham, receiver
1914–1919	Basingstoke Canal Syndicate Ltd	William Carter, mortgagee
1919–1923	in liquidation	W.H.C. Curtis, receiver
1923–1937	A.J. Harmsworth	
1937–1949	Weybridge, Woking & Aldershot Canal Co.	A.J. Harmsworth, chairman
1950–1973/6	New Basingstoke Canal Co.	S.E. Cooke, chairman
1973 to date	Hampshire County Council	(Hampshire section only)
1976 to date	Surrey County Council	(Surrey section only)

4. Officers and Servants of the Basingstoke Canal Navigation Company, 1778–1866

Chairman

1778–1796	Not appointed
1796–1816	Dr Robert Bland
1816–1832	Richard Birnie (knighted in 1821)
1832–1840	John Sloper
1840–1844	Uncertain
1844–1866	Peter Davey

Chairman Committee of Accounts

1787–1796	George Stubbs

Superintendant

1794–1797	George Stubbs

Surveyor & Engineer

1788–1790	William Jessop (consultant)
1788–1790	William Wright (resident)
1790–1793	Henry Eastburn (resident)
1793–1803	George Smith (resident)
1803–1813	Richard Allen (resident)
1829–1831	Francis Giles (consultant)
1849–1850	Martin (consultant)

Clerk

1778–1816	Charles Best
1816–1827	John Richard Birnie
1827–1866	Charles Headeach

Deputy Clerk

1799–1803	William Harrison (London)
1804–1806	Peter Jolit (London)
1806–1816	John Richard Birnie (Basingstoke)

Wharfinger (Basingstoke)

1794–1823	Thomas Adams
1823–1828	James Webb
1828–18??	Duties undertaken by clerk
1856–1866	Frederick W. Bushell (also traffic manager)

Wharfinger (Odiham)

1793–*c.* 1818	George Webb

5. BASINGSTOKE CANAL TRAFFIC RETURNS, 1791–1949

Year	*Tolls (£)*	*Tonnage passing on to or from the Wey Navigation*	*Total traffic (tons)*	*Balance of revenue over expenditure (£)*	*Source*
1791	1*	28	–	–	B
1792	25*	173	–	–	B
1793	170*	857	–	–	B
1794	1,000*	5,797	–	–	B
1795	2,165*	11,719	–	–	B
1796	2,750*	14,700*	–	–	B
1796–7	2,700*	14,300*	–	1,200 (loss)	B
1797–8	3,000*	13,466	–	614	B
1798–9	2,411*	11,639	–	626	B
1799–1800	3,515	17,877	–	1,345	B
1800–1	3,862	18,638	–	2,038	B
1801–2	3,814	18,737	–	1,925	B
1802–3	2,907	14,808	–	1,460	B
1803–4	2,877	12,077	–	373	B
1804–5	3,150*	15,000*	–	700*	B & M
1805–6	3,500*	17,000*	–	1,500*	B & M
1806–7	?	13,355	–	?	M
1807–8	?	15,232	–	?	M
1808–9	4,000*	16,810	–	1,700*	B & M
1809–10	4,000*	18,160	–	1,700*	B & M
1810–11	?	23,809	–	?	M
1811–12	?	22,249	–	?	M
1812–13	4,258	23,725	–	1,906	B
1813–14	3,486	16,305	–	813	B
1814–15	4,536	21,695	–	1,267	B
1815–16	4,033	20,106	–	1,511	B
1816–17	3,470	17,911	–	1,190	B
1817–18	3,350	18,240	–	1,503	B
1818–19	4,121	20,633	–	2,024	B
1819–20	4,400*	22,000*	–	2,000*	B
1820–1	3,898	20,050	–	1,784	B
1821–2	2,914	14,814	–	950	B
1822–3	2,394	11,971	–	626	B
1823–4	2,940	15,871	–	1,151	B
1824–5	3,250*	17,082	–	1,200*	B

* Some months averaged.

Year	Tolls (£)	Tonnage passing on to or from the Wey Navigation	Total traffic (tons)	Balance of revenue over expenditure (£)	Source
1825–6	3,001	15,258	–	1,002	B
1826–7	3,000*	14,573	–	1,200*	B
1827–8	3,528	18,428	–	1,621	B
1828–9	3,513	18,616	–	1,671	B
1829	3,292	19,384	–	455 (loss)	B
1830	3,295	19,845	–	419	B
1831	–	–	–	–	
1832	–	–	–	–	
1833	–	–	–	–	
1834	–	–	–	–	
1835	–	23,937	–	–	W
1835–6	4,875	25,988	32,000*	800	B & W
1836	5,000*	27,435	33,809	2,000*	B & W
1837	–	26,037	–	–	W
1838	–	33,879	–	–	W
1838–9	5,416	33,717	39,000*	2,500*	B & W
1839	–	25,622	–	–	W
1839–40	3,763	25,053	26,965	1,000	B & W
1840	–	24,877	–	–	W
1841	–	21,723	–	–	W
1842	–	21,553	–	–	W
1843	–	20,530	–	–	W
1844	–	17,916	–	–	W
1845	–	22,906	–	–	W
1846	–	16,200	–	–	W
1847	–	15,715	–	–	W
1848	–	16,212	22,364	–	W
1848–9	1,870	15,244	21,376	–	B & W
1849	–	11,202	–	–	W
1849–50	1,237	9,896	13,343	26	B & W
1850	–	10,167	–	–	W
1851	–	8,696	–	–	W
1852	–	8,666	–	–	W
1853	–	11,031	–	–	W
1854	–	10,669	–	–	W
1855	–	19,720	–	–	W
1856	–	27,664	32,486	–	W & b

* Some months averaged.

Year	*Tolls (£)*	*Tonnage passing on to or from the Wey Navigation*	*Total traffic (tons)*	*Balance of revenue over expenditure (£)*	*Source*
1857	–	19,720	24,595	–	W & b
1858	–	14,293	17,936	–	W & b
1859	–	15,371	19,165	–	W & b
1860	–	26,087	32,459	–	W & b
1861	1,005†	22,197	27,283	–	W & b
1862	877†	25,127	29,526	–	W & b
1863	–	24,795	29,978	–	W & b
1864	–	23,267	28,098	–	W & b
1864–5	1,610	22,899	33,743	–	B
1865	–	14,616	18,155	–	W & b
1865–6	1,212	12,925	20,598	60 (loss)	B
1866	–	12,553	15,850	–	W & b
1867	–	11,732	14,634	–	W & b
1868	–	12,126	–	–	W
1869	–	9,469	–	–	W
1870	–	8,632	–	–	W
1871	–	7,091	–	–	W
1872	–	5,829	–	–	W
1873	–	7,568	–	–	W
1874	–	8,296	–	–	W
1875	–	4,570	–	–	W
1876	–	4,435	–	–	W
1877	–	3,804	–	–	W
1878	–	3,415	–	–	W
1879	–	2,876	–	–	W
1880	–	2,296	–	–	W
1881	–	4,933	–	–	W
1882	–	645	–	–	W
1883	–	720	–	–	W
1884	–	1,864	–	–	W
1885	–	301	–	–	W
1886	–	1,246	–	–	W
1887	–	2,781	–	–	W
1888	1,346	3,623	4,187	nil	W & BoT
1889	–	3,922	–	–	W
1890	-	3,322	–	–	W
1891	–	4,375	–	–	W

† Six months only.

Year	*Tolls (£)*	*Tonnage passing on to or from the Wey Navigation*	*Total traffic (tons)*	*Balance of revenue over expenditure (£)*	*Source*
1892	–	3,021	–	–	W
1893	–	3,807	–	–	W
1894	–	5,152	–	–	W
1895	–	2,346	3,000*	–	W
1896	–	5,080	9,000*	–	W
1897	–	5,389	10,000*	–	W
1898	3,306	8,342	20,770	3,077 (loss)	W & BoT
1899	–	7,061	16,500*	–	W
1900	–	5,782	14,000*	–	W
1901	–	3,332	–	–	W
1902	–	1,766	–	–	W
1903	–	2,207	–	–	W
1904	–	1,984	–	–	W
1905	–	3,544	–	–	W
1906	–	3,519	–	–	W
1907	–	1,295	–	–	W
1908	–	2,131	–	–	W
1909	–	6,954	–	–	W
1910	–	9,394	–	–	W
1911	–	10,372	–	–	W
1912	–	9,296	–	–	W
1913	–	13,776	–	–	W
1914	–	12,790	–	–	W
1915	–	11,622	–	–	W
1916	–	14,154	–	–	W
1917	–	12,531	–	–	W
1918	–	17,942	–	–	W
1919	–	12,925	–	–	W
1920	–	13,836	–	–	W
1921	–	9,509	–	–	W
1922	–	15,644	–	–	W
1923	–	18,246	–	–	W
1924	–	17,182	–	–	W
1925	–	15,314	–	–	W
1926	–	16,647	–	–	W
1927	–	21,077	–	–	W
1928	–	19,026	–	–	W

* Some months averaged.

Year	*Tolls (£)*	*Tonnage passing on to or from the Wey Navigation*	*Total traffic (tons)*	*Balance of revenue over expenditure (£)*	*Source*
1929	–	18,506	–	–	W
1930	–	22,679	–	–	W
1931	–	20,092	–	–	W
1932	–	24,971	–	–	W
1933	–	28,526	–	–	W
1934	–	30,330	–	–	W
1935	–	31,577	–	–	W
1936	–	27,021	–	–	W
1937	–	17,387	–	–	W
1938	–	10,895	–	–	W
1939	–	8,931	–	–	W
1940	–	2,828	–	–	W
1941	–	5,159	–	–	W
1942	–	5,175	–	–	W
1943	–	3,934	–	–	W
1944	–	2,694	–	–	W
1945	–	4,171	–	–	W
1946	–	1,107	–	–	W
1947	–	1,689	–	–	W
1948	–	221	–	–	W
1949 (27 June)	–	130	–	–	W

Notes – The Wey Navigation traditionally calculated its traffic in loads of 25 cwt; there is uncertainty, however, in whether this was always done, for the figures contained in the Basingstoke Canal Navigation Company reports were in tons and in some instances they suggest that both navigations used the same unit of weight. I have therefore only converted loads into tons from 1890 onwards.

The Basingstoke Canal Company also altered the methods of calculating its tonnage in 1829. Before this date no account was taken of the carriage of chalk; by the old reckoning the figure for 1829 would have been 16,624. Between 1796 and 1806 chalk from the company's land was carried toll-free in return for £1,000 and £300 p.a. Local traffic was negligible after 1900.

The balance of revenue over expenditure excludes payments on the bond-debt. Where two years are given, the year ended on 25 March.

Sources – 1791–1866 Basingstoke Canal Navigation Reports (where available) (B)
1796–1812 Manning & Bray. *History of Surrey*, Vol. III, p. lix (M)
1835–1949 Wey Navigation Abstract of Accounts (W)
1856–1867 Basingstoke Canal Traffic Ledger (b)
1888 & 1898 Board of Trade Returns

Due to different methods of calculation, some of the figures from these sources contradict each other.

Notes

Abbreviations

BCN Basingstoke Canal News published by the Surrey & Hampshire Canal Society. Nos 1 to 129 were called Newsletter.

BCR Basingstoke Canal Navigation Company Report. These printed reports are held either at the Public Record Office, Kew or at the County Record Office, Winchester.

HRO Hampshire County Record Office, Winchester.

ICE Institution of Civil Engineers, Westminster.

JHC Journal of the House of Commons.

PRO Public Record Office.

1. Defoe, *A Tour through the Whole Island of Great Britain*, 1724, letter III, p. 15.
2. *Victoria County History: Hampshire & Isle of Wight*, Vol. V, p. 433.
3. Mudie, *Hampshire*, 1838, Vol. II, p. 37.
4. Celia Fiennes, *Through England on a Side-Saddle in the Time of William & Mary*, 1888, p. 20.
5. Baigent & Millard, *History of Basingstoke*, 1889, p. 555.
6. *Journal of the House of Lords*, xi, 675.
7. *Gentleman's Magazine*, 1778, p. 172; also *JHC*, 21 February 1771.
8. *JHC*, 23 February 1778.
9. J. Phillips, *A General History of Inland Navigation, Foreign and Domestic*, 1795, p. 232.
10. Plan at Hampshire County Record Office, Winchester.
11. *JHC*, 27 November, 5 December, 18 December 1770; 22 January, 23 January 1771.
12. *JHC*, 23 January, 31 January, 6 February 1771.
13. Jackman, *The Development of Transportation in Modern England*, 1916, Vol.I, p. 378.
14. *JHC*, 23 February 1778. Evidence of William Hodgkinson.
15. Pamphlet 'Proposals for a Navigable Canal from Basingstoke to the River Wey', 1777.
16. HRO 44m 69 G/1/1a.
17. *Gentleman's Magazine*, April 1778, Vol. 48, p. 172.
18. *JHC*, 23 February 1778.
19. *JHC*, 23 February 1778. Evidence of Thomas May.
20. Jackman, *op. cit.*, Vol. I, p. 398.
21. *JHC*, 15 April 1778.
22. *Ibid.*, 13 April 1778.
23. *The London Chronicle*, 7 March 1788.
24. *Observations on a scheme for extending the navigation of rivers Kennet & Avon so as to form a direct Inland communication between London, Bristol and the West of England, by canal from Newbury to Bath*, 1788, p. 13.
25. MS. Baxter to Portal, 15 December 1787 (HRO 5M52/TR3/20).
26. *Hampshire Chronicle*, 6 October 1788.
27. L.T.C. Rolt, *Great Engineers*, 1962, pp. 44–6.

28. Essex Record Office, D/DQs 135/2 and D/DRa 016. John Boyes & Ronald Russell, *The Canals of Eastern England*, 1977, p. 70.
29. Benjamin Henry Latrobe, *Talbot Hamlin* (O.U.P. New York) 1955, p. 27.
30. Humphrey Household, *The Thames & Severn Canal*, 1969, pp. 55–61.
31. *BCR*, 31 August 1789.
32. *BCR*, 23 February 1789.
33. *Reading Mercury & Oxford Gazette*, 22 December 1788.
34. Quoted by Charles Hadfield in *Canals of Southern England*, 1955, p. 77.
35. *London Chronicle*, 29 December 1788.
36. *Reading Mercury*, 24 November 1788.
37. Shaw, *A Tour of the West of England in 1788*, 1789, p. 545.
38. I am indebted to Mr Anthony Harmsworth for this information.
39. *BCR*, 22 February 1790.
40. Gotelee, *A History of Odiham*, 1932, p. 39.
41. Freeling, *Guide to the London & Southampton Railway*, 1839, p. 79.
42. *The Times*, 30 May 1963.
43. W.J. Davies, *The Nineteenth Century Token Coinage of Great Britain, Ireland, the Channel Islands and the Isle of Man*, p. xvi.
44. Letter from G. Stubbs to Lord Portmore and B. Langton, dated 16 May 1792 (Guildford Archivist's Office).
45. Basingstoke Canal Navigation Report, dated 27 August 1792.
46. *Hampshire Chronicle*, 26 November 1792.
47. *JHC*, 4 March 1793.
48. *Hampshire Observer*, 5 May 1794.
49. *Ibid.*, 9 June 1794.
50. *Reading Mercury & Oxford Gazette*, 22 August 1794 and *Hampshire Chronicle*, 25 August 1794.
51. Defoe, *A Tour through the Whole Island of Great Britain*, letter II, p. 88.
52. Shaw, *A Journey to the West of England in 1788*.
53. *BCR*, 17 October 1794; the copy in the Birmingham Reference Library lacks p. 3.
54. MSS. W. Alladay to G. Stubbs, 22 December 1794 (Wey Navigation Records). Alladay was lock-keeper at Thames Lock, Weybridge, 1779–1870.
55. *Reading Mercury & Oxford Gazette*, 30 May 1796.
56. Geo. III c.75, p. 137.
57. *BCR*, 10 October 1800.
58. *BCR*, 30 September 1795.
59. Printed extracts from Mr Stubbs's report dated 27 August 1796. (HRO)
60. Circular letter to shareholders dated 25 March 1797. (HRO)
61. *BCR*, 20 July 1800.
62. *Ibid.*, 10 October 1800.
63. *Dictionary of National Biography*, Vol. II, p. 661 gives his date of birth incorrectly as 1730. He was born in 1740 (see Carlisle, *Collections for a History of the Ancient Family of Bland*, 1826).
64. *BCR*, 21 September 1798.
65. *BCR*, 16 February 1804.
66. *BCR*, 12 February 1807.

67. *Ibid.*, 14 February 1805.
68. *An Address to the Public on the Basingstoke Canal Navigation*, 1783.
69. *BCR*, 1 June 1789.
70. MS. Rennie – Jessop, 30 December 1789 (ICE).
71. Report on the Extension of the Basingstoke Canal to Salisbury, London, 23 September 1790 (ICE).
72. Phillips, *General History of Inland Navigation* (4th edition), 1803.
73. Revd Arthur Young, *General View of the Agriculture of the County of Sussex*, 1808, p. 422. See also P.A.L. Vine, *London's Lost Route to Midhurst*, chapter VI.
74. Hadfield, *op. cit.*, p. 90.
75. *BCR* 20 July 1800, p. 4 (£16,000) and Printed MS George Stubbs to the Proprietors of the Basingstoke Canal Navigation, 18 August 1800, p. 2 which quotes Eastburn's estimate made in February 1793 of £20,000.
76. *Ibid.*, 16 August 1800.
77. Plan dated 30 September 1801 at Surrey Archivist's Office, Kingston-on-Thames.
78. W. Meyor, *A General View of the Agriculture of Berkshire*, 1809.
79. *BCR*, 10 February 1803, p. 2.
80. *Considerations on the intended Junction of the Ports of London, Southampton and Portsmouth by uniting the Basingstoke Canal (or the River Wey, at Godalming) with the River Itchen.*
81. *BCR*, 11 May 1809.
82. *BCR*, 9 February 1809; also Hadfield, *op. cit.*, 1955, p. 91.
83. A detailed account of the building and use of this line of navigation can be found in Vine, *London's Lost Route to the Sea*, 1965.
84. Kennet & Avon Canal Navigation Minutes, 7 February 1794, refer to a letter from Mr Best of Basingstoke offering the proprietors of the Kennet & Avon 330 shares in a canal proposed to be made from Hampstead Marshall near Newbury to Basingstoke.
85. HRO SM52/TR3/22.
86. *BCR*, 18 October 1810.
87. Report of the Kennet & Avon Committee of Management to the subscribers to the expenses of making surveys to find out the best line of navigation from the Kennet & Avon to London, 9 October 1810.
88. *Two reports of the Commissioners of the Thames Navigation on the objects and consequences of the several projected canals which interfere with the interests of that river; and on the present sufficient and still improving state of its Navigation*, published 1811 by order of the General Meeting at Oxford, 29 December 1810.
89. Remarks on a letter from Mr Page, London, 20 December 1810.
90. *An authentic description of the Kennet & Avon Canal to which are added observations upon the present state of the inland navigation of the south-western counties of England and of the counties of Monmouth, Glamorgan and Brecon in South Wales*, 1811.
91. Kennet & Avon Canal Navigation Report, 17 July 1832.
92. *Thames Navigation Observations upon the Evidence adduced before the Committee of the House of Commons upon the late application to Parliament for a Bill for making a navigable canal from the river Kennet, at Midgham, to join the Basingstoke Canal, to be called the Hants & Berks Canal*, 1825, p. 36.
93. *BCR*, 23 June 1825.

94. Kennet & Avon CNR, 28 June 1825.
95. *BCR*, 22 October 1827.
96. Kennet & Avon CNR, 4 December 1827.
97. *Hampshire Telegraph*, 25 May 1829 and *Derby Mercury*, 10 June 1829, p. 3.
98. Mudie, *Hampshire*, 1838, Vol. II, p. 38.
99. Kennet & Avon CNR, 19 January 1825.
100. *Victoria County History: Hampshire*, 1911, Vol. IV, p. 87.
101. *Annual Register of the Year 1820*, p. 52.
102. *BCR*, 21 October 1816. Dr Bland had previously announced that due to the death of two traders and the failure of two others, the company stood to lose £800 or £900.
103. MSS. Richard Birnie to Messrs Warne & Lewes of Basingstoke, 16 May 1818 (HCRO).
104. The diary rests at present in the Hampshire County Record Office at Winchester. Attwood died on 18 February 1870, aged 77.
105. See Vine, *London's Lost Route to the Sea*, 1965, p. 161.
106. George Bourne (pseud) William Smith, 1790–1858 potter and farmer, 1920.
107. *Victoria County History: Hampshire.*
108. Diary of Samuel Attwood, 29 April 1832.
109. Fay, *A Royal Road*, 1883, p. 10.
110. J. Francis, *History of the English Railway*, 1851.
111. Williams, *London & South-Western Railway*, Vol. I, 1968.
112. Freeling, *London & Southampton Railway Companion*, 1839, p. 62.
113. Francis Wishaw, *Railways of Great Britain & Ireland*, 1840, p. 295.
114. Circular letter from the Clerk of the Basingstoke Canal Company to bond-holders dated 8 August 1836.
115. *Pontet* v. *Basingstoke Canal Co.*, 3 Bingham New Cases, p. 433, et seq. (1837).
116. *BCR*, 1 June 1840.
117. *BCR*, 4 November 1839.
118. Charles Hadfield, *op. cit.*, 1955, p. 270.
119. 'The Advantages and Profits of the London & Southampton Railway, analysed', quoted by C.F. Dendy-Marshall, *History of the Southern Railway*, 1936.
120. *The Journey Book of England, Hampshire*, 1841, p. 6.
121. Jackman, *The Development of Transportation in Modern England*, 1916, Vol. II, p. 494.
122. Second report of the Select Committee on the Amalgamation of Railways and Canals, 1846, Parliamentary Papers, Vol. XIII, pp. 95–6.
123. Corporation of London Navigtion Committee Minutes, 28 November 1844.
124. *BCR*, 3 June 1850.
125. Adrian Gray, *South-Eastern Railway*, 1990, p. 182.
126. PRO: War Office Correspondence Out-Letters (W.O.3.) 326.
127. Lord Hardinge wrote to the Prince Consort on 5 December 1853 stating that he hoped shortly to report progress, 'knowing the very deep and successful interest which your Royal Highness takes in the matter, and how much the Army will be indebted to your Royal Highness for this permanent and admirable camp of instructions'.
128. Cole, *The Story of Aldershot*, 1951, p. 31.
129. *Ibid.*, pp. 39–40.
130. Corporation of London Navigation Committee Minutes, 12 February 1855.

131. MSS. Stephen William Leach to Worshipful Committee of the Thames Navigation, 5 July 1855, quoted in Corporation of London Navigation Committee Minutes.
132. Mrs Young, Aldershot and All About It, 1857, p. 91.
133. Published in *All the Year Round*, which was edited by Charles Dickens.
134. *History, Gazetteer and Directory of Hampshire*, 1859, p. 486.
135. *The Times*, 5 November 1861.
136. Wey Navigation Records, Guildford.
137. *The Times*, 5 November 1861.
138. *The Times*, 19 December 1865.
139. Reprinted in *The Times*, 26 March 1866.
140. *County Chronicle & Weekly Advertiser*, 10 and 24 November 1818.
141. London & Southampton Railway Bill. Minutes of Evidence – Samuel Jones, 23 June 1834. See also a letter to the editor re railways in *Bristol Journal,* 4 January 1834, which said the shares stood at 105*s*.
142. *BCR*, 13 February 1806.
143. *The Times*, 5 November 1861, reports that a payment of 2 per cent was to be made on the bond-debt in December.
144. London & Southampton Railway Bill. Minutes of Evidence 27 May 1834.
145. Taunt, *Illustrated Map of the Thames* (3rd edition), 1879, p. 83.
146. *The Times*, 29 November 1913.
147. Auction particulars, July 1883. Conditions of Sale, para. 9.
148. Williams, *The London & South-Western Railway*, Vol. II, 1973, p. 80.
149. *Douglas* v. *Leeming* 1882 D.2144 (High Court of Justice, Chancery Division).
150. PRO BT 31/3231/18917.
151. *H.C.J. Douglas* v. *Barrett* 1887 D no. 73.
152. Report Colonel F.H. Rich to BoT Railway Dept, 22 April 1891.
153. PRO office index 48639. File index BT 31/6914. By December 1897 Sir Frederick's holding had been reduced to 2,000 shares.
154. MSS. J. Melland-Smith to W. Stevens, 14 April 1896.
155. PRO BT 34/96921/2202 and BT 31/96921/12294.
156. PRO BT 31/7607/54309 contains a plan showing that the cut was not constructed before 1897.
157. In the private possessions of the late Mr Alec Harmsworth's executors.
158. Letter from the receiver of the Hampshire Brick & Tile Company to the Wey Navigation dated 25 October 1901.
159. The South-Western Railway (Various powers) Act, 1897 Section 5 (6).
160. MSS. Particulars of Woking, Aldershot & Basingstoke Canal & Navigation Co. Ltd, 1900.
161. *History of the Southern Railway*, 1937, chapter XI.
162. Prothero, & Clarke, *A New Oarsman's Guide to the Rivers and Canals of Great Britain and Ireland*, 1896, pp. 11–13.
163. *South Western Gazette*, 1 December 1904.
164. S.T. Felstead, *Horatio Bottomley – A Biography of an Outstanding Personality*, 1936, pp. 103–7.
165. PRO BT 31/96921/12294 and BT 34/96921/2202.
166. Julian Symons, *Horatio Bottomley*, 1955, pp. 112–13.
167. *Daily Telegraph*, 16 October 1909.

168. Woking Urban District Council (Basingstoke Canal) Act 1911, preamble.
169. *The Times*, 19 February 1912.
170. *Woking Observer & Weybridge Chronicle*, 5 November 1913.
171. *The Times*, 29 November 1913.
172. MSS. Messrs Fraser & Christian to A.J. Harmsworth, 28 November 1913.
173. Leslie Thomas (b. 1914) from *Landmarks, a Book of Topographical Verse for England and Wales* (Cambridge University Press), 1943, p. 56.
174. Last ledger entry, 22 May 1904 referred to the passage of the empty *Crookham* from the Wey to Basingstoke. The last meeting of the Hampshire Brick & Tile Co. was held by the receiver in 1906, but no quorum attended.
175. *The Times*, 1 June 1914. The barge in fact only reached Basing Wharf. *The Times* stated that 'in February Mr A. Harmsworth of Ash Vale proved that it was theoretically possible'.
176. *Daily Mirror*, 18, 19 and 20 November 1913; *Weekly Dispatch*, 16 November 1913; *Daily Express*, 18 and 19 November, 9 and 10 December 1913; *Hampshire Observer*, 13 December 1913. Gaumont-Graphic News filmed the task of filling a pound, opening a swing-bridge and man-hauling the barge through the weeds at Mapledurwell.
177. J.R. Colville, *Man of Valour*, 1972, p. 18.
178. 17 & 18 Geo. V c.13, Section II. This section has not been repealed by any of the subsequent Acts.
179. *The Times*, 1 June 1914.
180. *Woking Observer & Weybridge Chronicle*, 6 May 1914.
181. C.S. Forester, *Long Before Forty*, 1967, pp. 84–5. In 'Hornblower and the Atropos', 1953, Forester uses his extensive knowledge of English canals to send his hero and family *en route* from Gloucester to London through the Thames & Severn Canal.
182. Romantic Friendship – The Letters of Cyril Connolly to Noel Blakiston 1973.
183. De Salis, *Bradshaw's Canals & Navigable Rivers of England & Wales*, 1928, p. 424.
184. Harmsworth, *Boats from the Basingstoke's Past*, 1969, p. 2 and p. 4.
185. Anon, *The English Scene (A&C Black)*, 1930, pp. 38–9.
186. M. Gotelee, *A History of Odiham*, 1932, p. 39. The tunnel remains navigable for 1,000 yards, but the crown of the roof has given way some 150 yards from the west entrance.
187. Martin Thornhill in *Explorer's Hampshire*, 1952, p. 48 refers to 'a primitive device being used of attaching wheels to the barges so that they could be drawn overland to the tunnel's other exit'. I can trace no evidence to support this story.
188. *The Times*, 13 December 1948.
189. Auction Particulars, 1 March 1949, general remarks.
190. Aickman, *The River Runs Uphill*, 1986, pp. 149–59; *BCN* 121, May 1985, pp. 4–5.
191. *The Windlass*, February 1962.
192. *The Windlass*, June 1962.
193. *Surrey Advertiser*, 13 May 1967.
194. *BCN* 26, November 1969.
195. *BCN* 39, July 1971.
196. MS. Roger Cansdale to the author, 30 June 1991.
197. *Basingstoke Canal: the Case for Restoration*, 1968, p. 51.
198. Dr R.E. Stebbings, *Bats in Greywell Tunnel*, June 1992.

Notes 163 to 199 are new to this edition.

BIBLIOGRAPHY

(I) ACTS OF PARLIAMENT

1778 An Act for making a navigable canal from the town of Basingstoke, in the county of Southampton, to communicate with the River Wey, in the parish of Chertsey, in the county of Surrey, and to the south-east side of the Turnpike road in the parish of Turgiss, in the said county of Southampton. (Basingstoke Canal Act.)

1793 An Act for effectually carrying into execution an Act of Parliament of the eighteenth year of his present Majesty, for making a navigable canal from the town of Basingstoke, in the county of Southampton, to communicate with the River Wey in the parish of Chertsey in the county of Surrey, and to the south-east side of the Turnpike road in the parish of Turgiss, in the said county of Southampton. (Basingstoke Canal Act.)

1894 An Act to confirm the provisional order made by the Board of Trade under 'The Railway & Canal Traffic Act 1888' containing the Classification of Merchandise Traffic, and the Schedule of Maximum Tolls and Charges applicable thereto, for the River Ancholme Navigation and certain other canals. (Canal Tolls and Charges Provisional Order No. 7 (River Ancholme, etc.) Order Confirmation Act.)

1911 An Act to authorize the urban district council of Woking to exercise some of the powers contained in the Act 18 Geo. III c.75 as regards certain bridges vested in the Company of Proprietors of the Basingstoke Canal Navigation, and for other purposes. (Woking UDC (Basingstoke Canal) Act.)

(II) BOOKS OF REFERENCE

1724 Defoe, Daniel, *A Tour through the Whole Island of Great Britain.*

1789 Shaw, Revd Stebbing, *A Journey to the West of England in 1788.*

1792 Phillips, J., *General History of Inland Navigation, Foreign and Domestic: containing a complete account of the canals already executed in England, with consideration on those projected.*

1794 Malcolm, W.J. & J., *A General View of the Agriculture of the County of Surrey.*

1798 Marshall, William, *The Rural Economy of the Southern Counties.*

1803 Phillips, J., *General History of Inland Navigation* (4th edition).

1804–14 Manning, Revd O. & Gray, W., *The History and Antiquities of the County of Surrey.*

1808 Anon, *Observations on the Proposed Junction Canal between Winchester and the Basingstoke Canal.*

1810 Allnutt, Zachariah, *Useful and Current Accounts of the Navigation of the Rivers and Canals West of London* (2nd edition).
Vancouver, Charles, *General View of the Agriculture of Hampshire.*
1813 Stevenson, W., *General View of the Agriculture of the County of Surrey.*
1817 Marshall, William, *Review of the Reports to the Board of Agriculture for the Southern and Peninsular Departments of England.*
1825 Thames Navigation Commissioners, *Observations upon the Evidence adduced before the Committee of the House of Commons upon the late application to Parliament for a Bill for making a navigable Canal from the river Kennet at Midgham to join the Basingstoke Canal, to be called the Hants & Berks Canal.*
Carlisle, Nicholas, *Collections for a History of the Ancient Family of Bland.*
1831 Priestly, Joseph, *Historical Account of the Navigable Rivers, Canals and Railways throughout Great Britain.*
1838 Mudie, Robert, *Hampshire: its past and present condition, and future prospects.*
1839 Freeling, Arthur, *Guide to the London & Southampton Railway.*
1857 *The Oarsman's Guide to the Thames and Other Rivers* (2nd edition).
Mrs Young, *Aldershot and All About It with Gossip, Literary, Military and Pictorial.*
1861–9 Woodward, B.B., Willis, T.C. & Lockhart, C., *General History of Hampshire.*
1879 Taunt, H.W., *Illustrated Map of the Thames* (3rd edition).
1883 Fay, Sam, *A Royal Road, being the History of the London & South-Western Railway from 1825 to the present time.*
1889 Baigent, F.J. & Millard, J.E., *History of the Ancient Town and Manor of Basingstoke in the County of Southampton.*
1890 Returns made to the Board of Trade in respect of the Canals and Navigations in the United Kingdom for the year 1888.
1896 Prothero, F.E. & Clarke, W.A., *A New Oarsman's Guide to the Rivers and Canals of Great Britain and Ireland.*
1897 De Salis, H.R., *Chronology of Inland Navigation.*
Taunt, H.W., *Illustrated Map of the Thames* (6th edition).
1899 Returns made to the Board of Trade in respect of the Canals and Navigations in the United Kingdom for the year 1898.
1909 Knight, J.H., *Reminiscences of a Country Town* (Farnham).
1912 *Victoria County History: Hampshire and Isle of Wight*, Vol. IV.
1916 Bonthron. P., *My Holidays on Inland Waterways.*
Jackman, W.T., *The Development of Transportation in Modern England.*
1928 De Salis, H.R., *Bradshaw's Canals & Navigable Rivers of England & Wales.*
1932 Gotelee, M., *A Record of the Past History of the Town, Park, Manor & Castle of Odiham.*
1949 Palmer, W.T., *Wanderings in Surrey.*
1950 Hadfield, Charles, *British Canals: an illustrated history.*
1951 Cole, Howard, *The Story of Aldershot.*
1965 Vine, P.A.L., *London's Lost Route to the Sea* – A historical account of the inland navigations which linked the Thames to the English Channel.
1966 Welch, Edwin, *The Bankrupt Canal. Southampton & Salisbury* 1795–1808.
1968 Clew, Kenneth R., *The Kennet & Avon Canal.*

	Jebens, D., *Basingstoke Canal – the Case for Restoration.*
	Vine, P.A.L., *London's Lost Route to Basingstoke* (1st edition).
1968–73	Williams, R.A., *The London & South-Western Railway*, Vols. I and II.
1969	Hadfield, Charles, *The Canals of South and South-East England.*
	Harmsworth, Tony, *Boats from the Basingstoke's Past.*
1973	Crocker, Glenys, *A History of the Basingstoke Canal.*
1977	Potter, R.W.F., *Hampshire Harvest.*
1982	Crosby, Alan, *A History of Woking.*
1984	Childerhouse, Tim, *Bygone Aldershot*
1985	Jebens, D. & Robinson, D., *Basingstoke Canal Restoration.*
1986	Aickman, Robert, *The River Runs Uphill.*
1987	Vine, P.A.L., *Surrey Waterways.*
1990	Vine, P.A.L., *Hampshire Waterways.*
	Surrey & Hampshire Canal Society, *The Promise of the Western End.*

(III) PERIODICALS

The Basingstoke Canal Newsletter published by the Surrey & Hampshire Canal Society provides current information and historical detail. The first news sheet appeared in January 1967, illustrations appeared in 1974, the centenary issue occurred in December 1981 and the summer issue for 1993 was no. 162.

'The Future of the Basingstoke Canal' (*The Motor Boat*, 1905).
'The Basingstoke Canal', Willans, Kyrle (*Country Life*, 24 December 1948).
'For Sale by Auction' (*Illustrated London News*, 19 February 1949).
'The Basingstoke Canal', Chaloner, W.H. and Mather, F.C. (*Edgar Allen News*, August 1949).
'Basingstoke? You can't get there by boat, can you?', Woolgar, F. (*Motor Boat & Yachting*, 15 December 1967).
'Future of the Basingstoke Canal – The Prospects of Revival', Vine, P.A.L. (*Country Life*, 6 February 1969).
'Battle for the Basingstoke', Palmer, Graham (*Narrow Boat*, December 1984).

INDEX